ROOFING READY RECK

METRIC AND IMPERIAL DIMENSIC
TIMBER ROOFS OF ANY
SPAN AND PITCH

ROOFING READY RECKONER

METRIC AND IMPERIAL DIMENSIONS FOR
TIMBER ROOFS OF ANY
SPAN AND PITCH

By Ralph Goss

Third edition revised by
Chris N. Mindham

Blackwell
Science

© 2001 by the estate of Ralph Goss and Blackwell Science Ltd

Blackwell Science Ltd, a Blackwell Publishing company
Editorial Offices:
Blackwell Science Ltd, 9600 Garsington Road, Oxford OX4 2DQ, UK
Tel: +44 (0) 1865 776868
Blackwell Publishing Inc., 350 Main Street, Malden, MA 02148-5020, USA
Tel: +1 781 388 8250
Blackwell Science Asia Pty, 550 Swanston Street, Carlton, Victoria 3053, Australia
Tel: +61 (0)3 8359 1011

The right of the Author to be identified as the Author of this Work has been asserted in accordance with the Copyright, Designs and Patents Act 1988.

All rights reserved. No part of this publication may be reproduced, stored in a retrieval system, or transmitted, in any form or by any means, electronic, mechanical, photocopying, recording or otherwise, except as permitted by the UK Copyright, Designs and Patents Act 1988, without the prior permission of the publisher.

Second edition published 1987
Third edition published 2001
Reprinted 2002, 2003, 2004, 2006

ISBN-10: 0-632-05765-3
ISBN-13: 978-0-632-05765-8

Library of Congress Cataloging-in-Publication Data
Roofing ready reckoner: metric and imperial dimensions for timber roofs of any span and pitch/Ralph Goss.—3rd ed./revised by Chris N. Mindham.
p. cm.
ISBN 0-632-05765-3 (alk. paper)
1. Roofs—Handbooks, manuals, etc.
2. Carpentry—Mathematics. 3. Roofing.
4. Engineering mathematics—Formulae.
I. Mindham, C.N. (Chris N.) II. Title.

TH2401 .G67 2001
695—dc21

00-140120

A catalogue record for this title is available from the British Library

Set in 9.5/11pt Univers
by DP Photosetting, Aylesbury, Bucks
Printed and bound in India
by Gopsons Papers Ltd, Noida

The publisher's policy is to use permanent paper from mills that operate a sustainable forestry policy, and which has been manufactured from pulp processed using acid-free and elementary chlorine-free practices. Furthermore, the publisher ensures that the text paper and cover board used have met acceptable environmental accreditation standards.

For further information on Blackwell Publishing, visit our website:
www.blackwellpublishing.com

CONTENTS

1 INTRODUCTION

The aim of this book is to provide a quick and easily usable reference for those constructing roofs for new buildings and extensions using cut roof or trussed rafter construction. The tables include data for common rafters, hips and valley and can also be used for attic construction. This third edition includes the $2\frac{1}{2}^\circ$ pitch increments used in trussed rafter roof construction for pitches from $17\frac{1}{2}^\circ$–$42\frac{1}{2}^\circ$. The data for steeper pitches and mansard roofs up to 75° is retained.

The addition of helpful illustrations and text on wall plate and gable end strapping, wind bracing, truss clips and other roofing metalwork plus valuable information on the tools and equipment to carry out the work, extends the guidance on roof construction.

Although the text is in metric units, imperial comparison is shown in brackets after each dimension with separate imperial worked examples. The all important data tables are retained in both metric and imperial units respecting the continuing popularity of this traditional method of measurement.

2 ROOFING TERMINOLOGY

Wall plate The 'foundation' of the roof usually 50 × 100 mm wide (2″ × 4″), must be bedded solid, level and straight on the top of the wall, or nailed to the timber framed panel and strapped in place to prevent movement from the structure.

Purlin Member carrying part load of the long common rafters, traditionally placed at right angles to the rafter but now more commonly fixed vertically.

Pitch The angle made by the slope of the roof with the horizontal. This may be stated in degrees on the drawing, or may have to be measured by protractor from the drawing, or may have to be calculated by measurement if the new work is to match an existing roof.

Ridge The timber at the top of the roof where the rafters meet, giving a longitudinal tie to the roof structure, commonly 38 mm ($1\frac{1}{2}''$) thick, and of a depth equal to the top cut on the rafter plus approximately 38 mm ($1\frac{1}{2}''$). This depth will depend upon the pitch of the roof and the tile batten thickness.

Common rafter The timber running from the ridge, down over the purlin if fitted, over the wall plate, and to the back of the fascia.

Jack rafter The timber running from the hip rafter down over the purlin if fitted, over the wall plate, and to the back of the fascia.

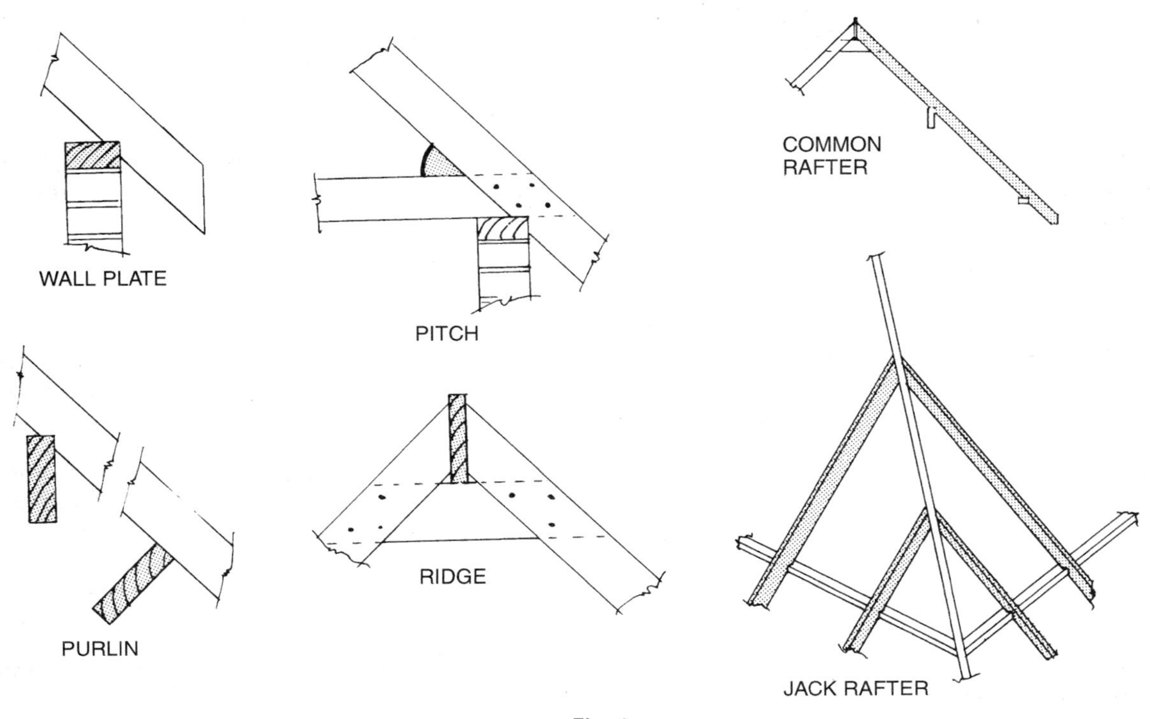

Fig. 1

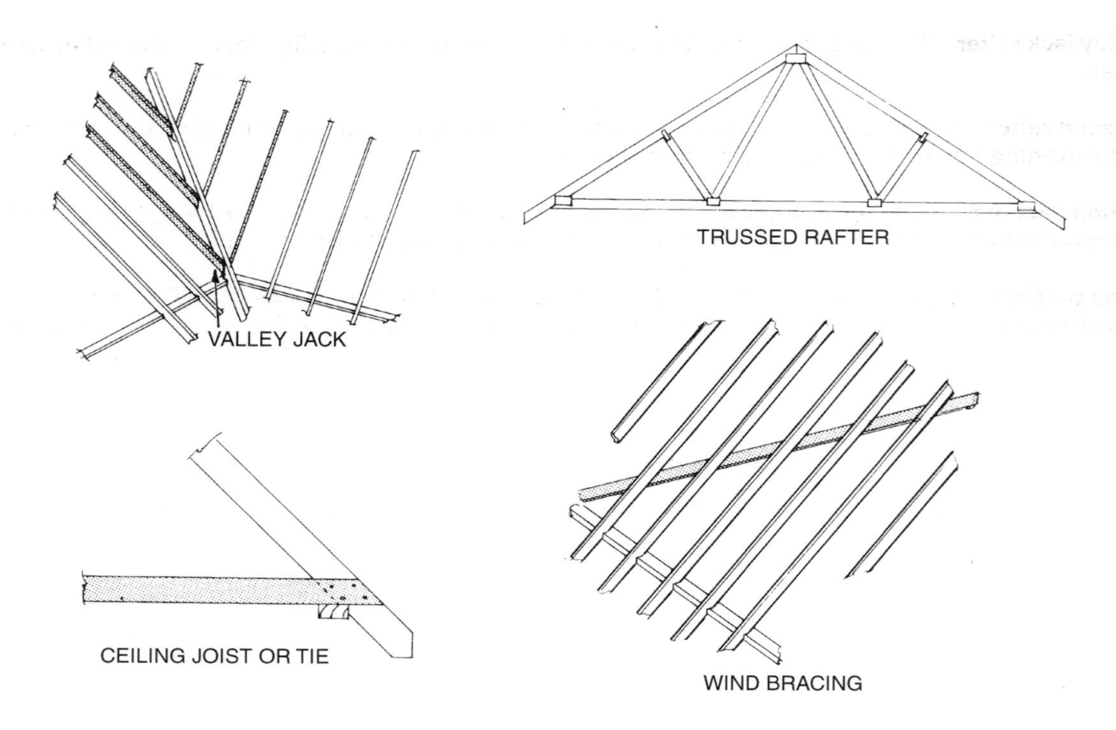
VALLEY JACK
TRUSSED RAFTER
CEILING JOIST OR TIE
WIND BRACING

Fig. 2.

Valley jack rafter The timber running from the ridge, down over the purlin, down to the valley board or rafter.

Trussed rafter A prefabricated framework incorporating rafter, ceiling joist (or tie), and strengthening webs forming a fully triangulated structural element.

Ceiling joist or tie Timber supporting the ceiling of the building, but often importantly 'tieing' the feet of the common and jack rafters together thus triangulating and stabilising the roof.

Wind bracing Usually 25 mm × 100 mm (1″ × 4″) timber nailed to the underside of rafters and trussed rafters running at approximately 45° to them, to triangulate and stabilise the roof in its vertical plain.

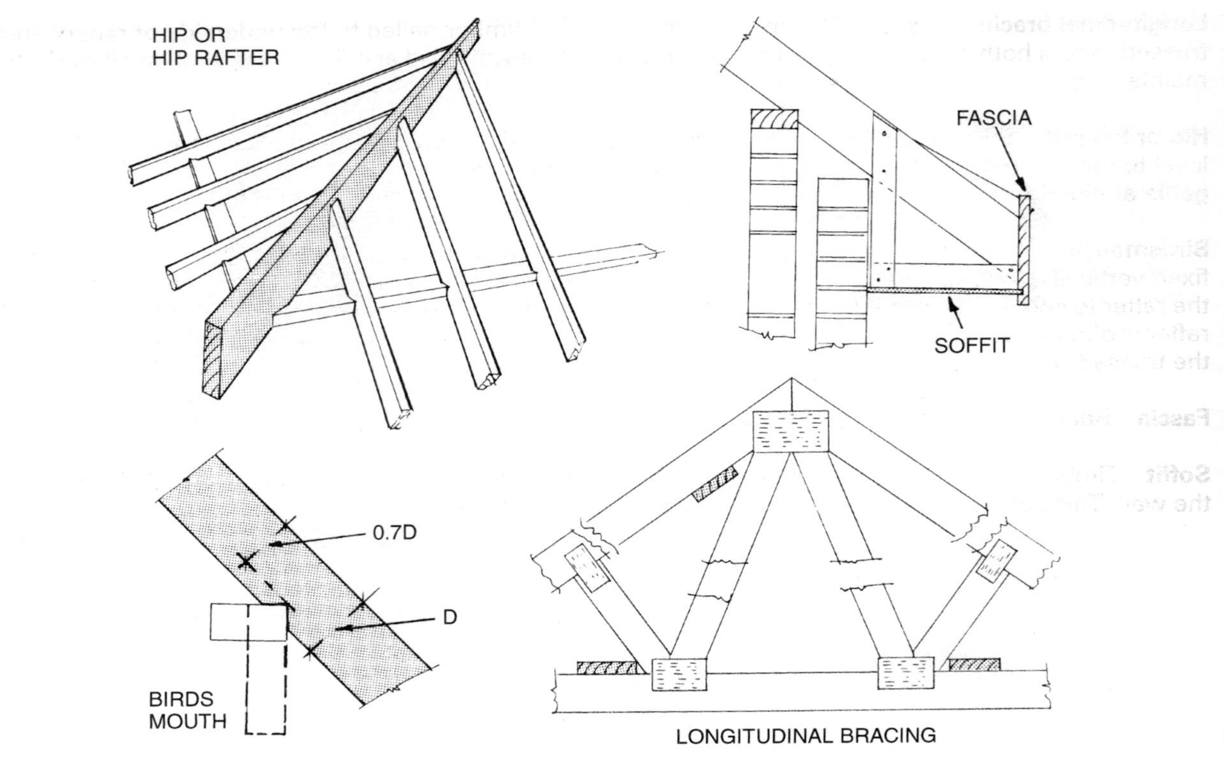
HIP OR
HIP RAFTER
FASCIA
SOFFIT
0.7D
D
BIRDS
MOUTH
LONGITUDINAL BRACING

Fig. 3.

Longitudinal bracing Usually 25 mm × 100 mm (1″ × 4″) timber nailed to the underside of rafters and trussed rafters both at the ridge position on a trussed rafter roof, and at ceiling joist level on all roofs, to maintain accurate spacing and stiffening of the members to which it is fixed.

Hip or hip rafter This is a substantial timber member running from the corner of the roof at wall plate level to the end of the ridge. In some designs the hip may stop lower down the roof producing a small gable at high level.

Birdsmouth The cut in rafters at the fixing point to the wall plate and or the purlin (where purlins are fixed vertically), this should leave at least 0.7 × the depth of the rafter to give the strength necessary for the rafter to continue to provide an over hang to the roof. If a common rafter is fitted as part of a trussed rafter roofing system, then the 0.7 × the depth of the rafter must be the same as the depth of the rafter on the trussed rafter component.

Fascia Board fixed to the rafter feet, supporting both gutter and soffit.

Soffit Timber board or sheet material used to close off the over hang between the back of the fascia and the wall. This soffit may have a roof ventilation system built into it.

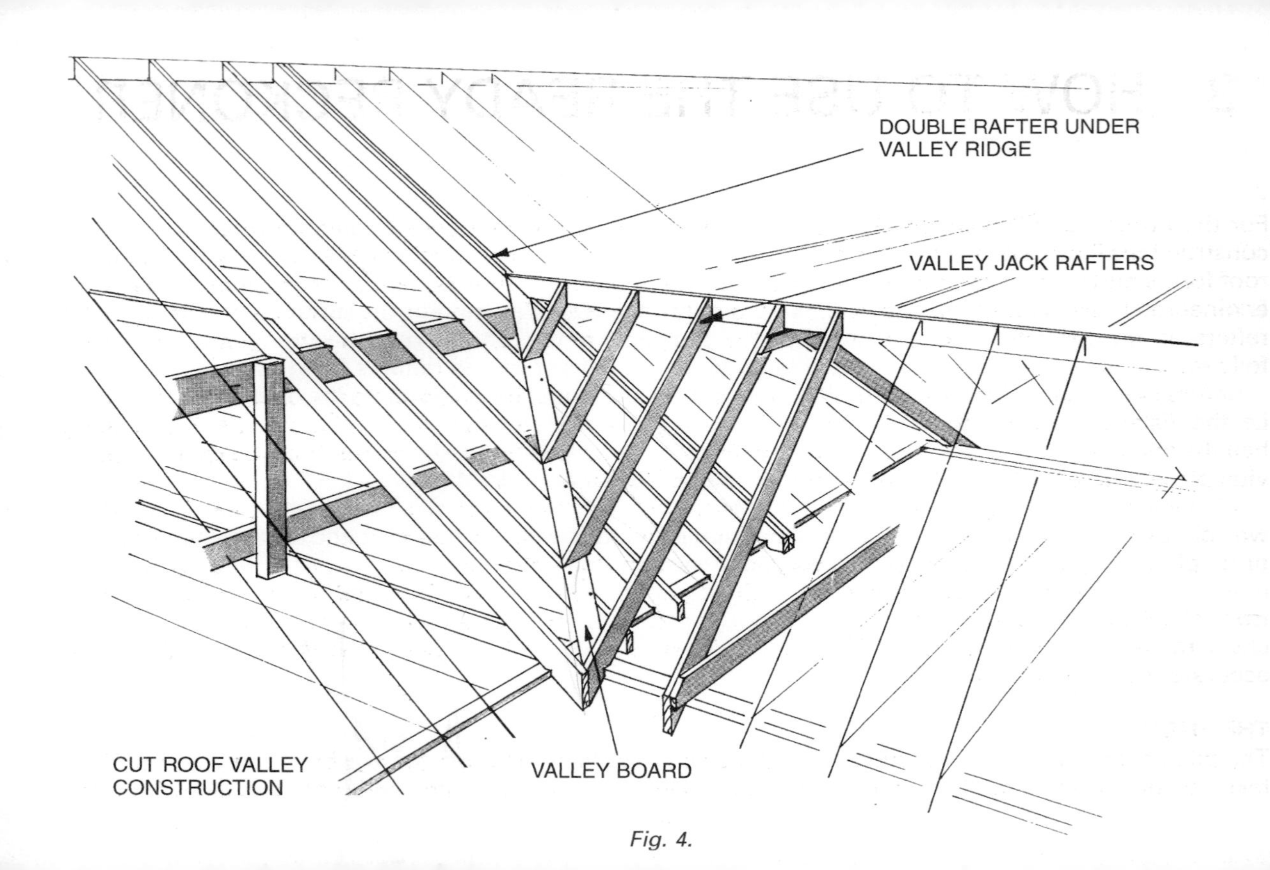
DOUBLE RAFTER UNDER
VALLEY RIDGE
VALLEY JACK RAFTERS
CUT ROOF VALLEY
CONSTRUCTION
VALLEY BOARD

Fig. 4.

3 HOW TO USE THE READY RECKONER

For the purposes of explaining the use of the ready reckoner, reference should be made to the roof constructions illustrated in Figs 4 and 5. In practical terms these constructions will cover most traditional roof forms and take account of the hip and valley infills used on trussed rafter construction unless fully engineered trussed rafter hip and valleys have been designed. The cutting angles on all timbers for infill rafters on trussed rafter roofs, especially attic designs, can be calculated using the data tables which follow.

Before cutting any of the roof timbers, two vital pieces of information must be known. Firstly the span i.e. the distance between the outer faces of the wall plate, and secondly the 'run' of the rafter, this being half the span assuming that it is an equally pitched roof with the ridge in the middle of the span. Another vital piece of information is the pitch or the 'rise' of the roof.

Whenever possible the carpenter who is to construct the roof should at least supervise the fixing of the wall plates. These must be straight, level and parallel to each other. Where the roof has to be fitted to a 'T' or 'L' plan form of building, then the carpenter should check that the wall plates of the projections to either side of the main roof are at a true right angle unless of course designed to be otherwise. Apart from checking overall dimensions with a steel tape, modern laser levels make it quick and simple to check the level of the plate very effectively, and it is this level of the wall plate which is so important to accurate roof construction.

THE PITCH

The pitch of the roof to be constructed should be clearly stated on the drawings but if not this should be taken by protractor from the drawings, possibly extending the ceiling line and rafter line away from the

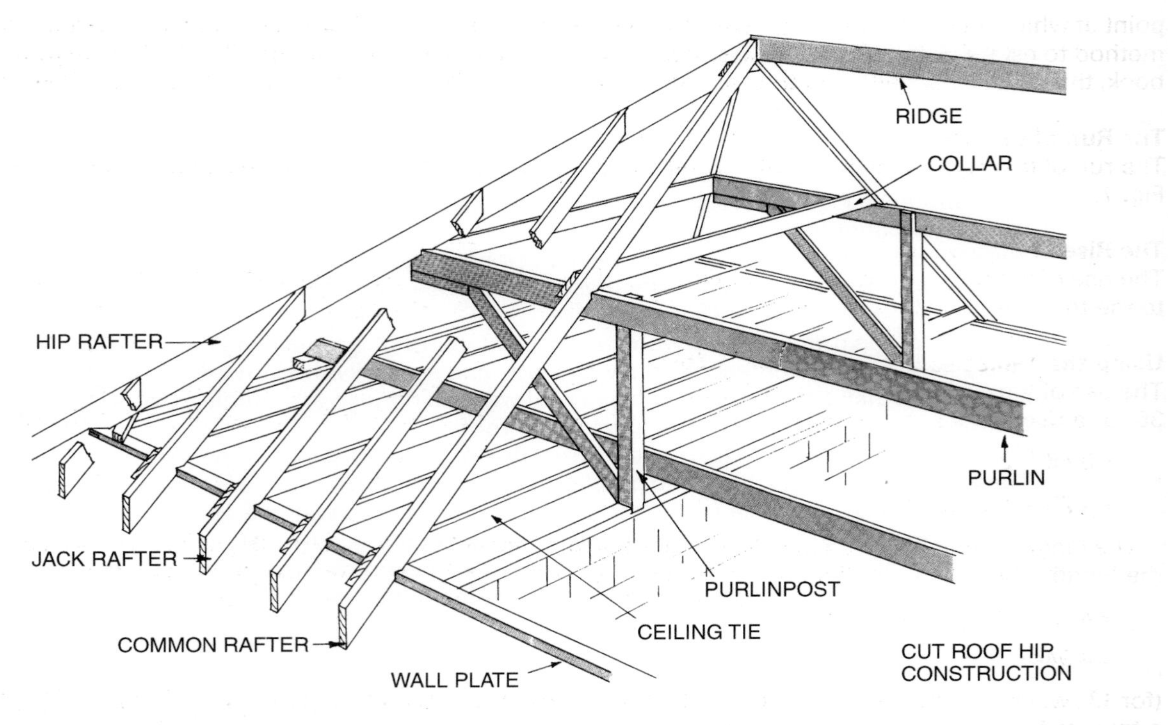
RIDGE
COLLAR
HIP RAFTER
PURLIN
JACK RAFTER
PURLINPOST
CEILING TIE
COMMON RAFTER
WALL PLATE
CUT ROOF HIP
CONSTRUCTION

Fig. 5.

point at which they meet making it easier to get an accurate reading from the protractor. An alternative method to establish the pitch is to calculate the rise of the roof per unit of 'run'. To use the tables in this book, this must be stated in metres rise per metre run or in feet and inches rise per foot run. See Fig. 6.

The Run of the Rafter

The run of the rafter is the horizontal distance covered by the rafter from the wall plate to the ridge. See Fig. 7.

The Rise of the Rafter

The rise of the rafter is the height from the top of the rafter vertically above the outside of the wall plate, to the top of the rafter at the centre line of the ridge position. See Fig. 7.

Using the Tables to Cut a Common Rafter

The use of the tables is best explained by a worked example and to do this we will take a roof of pitch at 36° or a rise of 0.727 m per metre run ($8\frac{3}{4}''$ per foot run), and a span of 8.46 m (27′ 9″). Then the run:

$$= 8.46 \div 2$$

$$= 4.23\ \text{m}\ (13'\ 10\tfrac{1}{2}'')$$

The length of the rafter can now be calculated from the tables referring to 36° pitch. It can be seen that the length of the rafter for 1 m of run = 1.236 m ($1.2\frac{7}{8}$ per 1′) therefore the length for 4 m of roof:

$$= 4 \times 1.236$$

$$= 4.944\ \text{m}$$

(for 13′, we must add from the tables the 10′ and 3′ run of rafter giving a total length of $12'4\frac{3}{8}'' + 3'8\frac{1}{2}''$ giving a total of $16'0\frac{7}{8}''$).

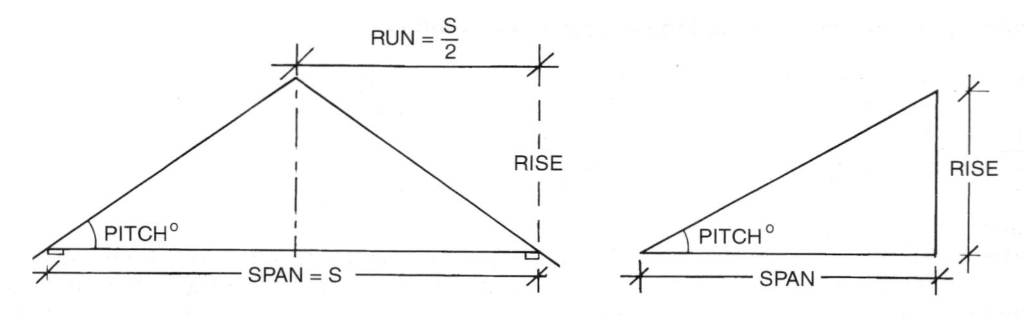

Fig. 6.

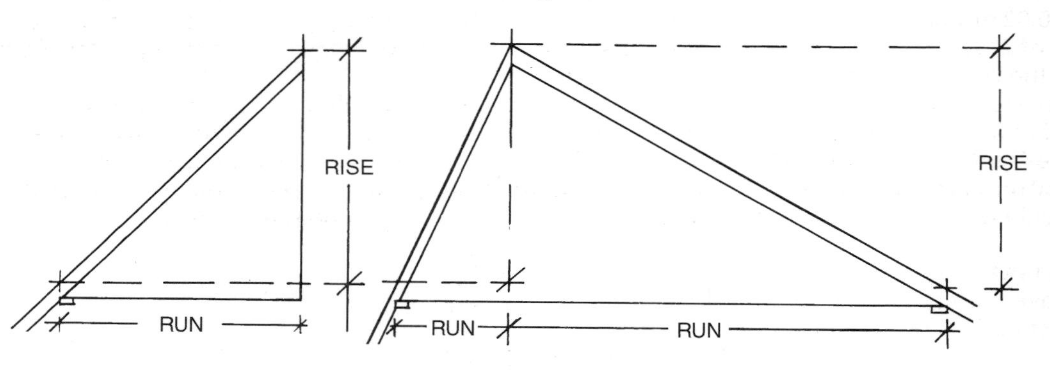

Fig. 7.

The calculation for the whole rafter length now looks as follows:

4 m	=	4.944 m	13′	=	16′0$\frac{7}{8}$″
0.2 m	=	0.247 m	10$\frac{1}{2}$″	=	1′1″
0.03 m	=	0.0371 m	–		–
4.23 m	=	5.2281 m	13′10$\frac{1}{2}$″	=	17′1$\frac{7}{8}$″

Now we have the basic length of the rafter.

To calculate the *exact* length of the rafter, the length above must be reduced by $\frac{1}{2}$ the thickness of the ridge, this can be calculated as above if perfection is required. For a ridge thickness of 40 mm, the length of run of the rafter must be reduced by 20 mm, this gives a reduction in rafter length from the tables of 0.0247 m or 24.7 mm. (Tables have to be modified by a factor of 10 because the run above is 20 mm which is 0.02 m and not 0.2 mm as illustrated in the tables. Similarly on imperial assuming a ridge thickness of 2″ which equals a reduction in rafter run of 1″, then by reference from the tables it can be seen that the rafter must be reduced by 1$\frac{1}{4}$″ in length).

We then need to add to the rafter the extra length needed to cover the over hang. This can be simply calculated in the same way by finding the length of over hang from the outside of the wall plate to the back of the fascia (see Fig. 8), and we will assume for the purposes of this calculation that this over hang gives a 450 mm run (18″), then again by reference to the tables, it will be seen that the additional rafter length required is 512 mm (1′ 10$\frac{1}{4}$″). This now gives an overall rafter length as follows:

Basic rafter		5.2281		17′1$\frac{7}{8}$″
Add over hang	+	0.556	+	1′10$\frac{1}{4}$″
Deduct half ridge	–	0.0247	–	1$\frac{1}{4}$″
		5.7594 m		18′10$\frac{7}{8}$″

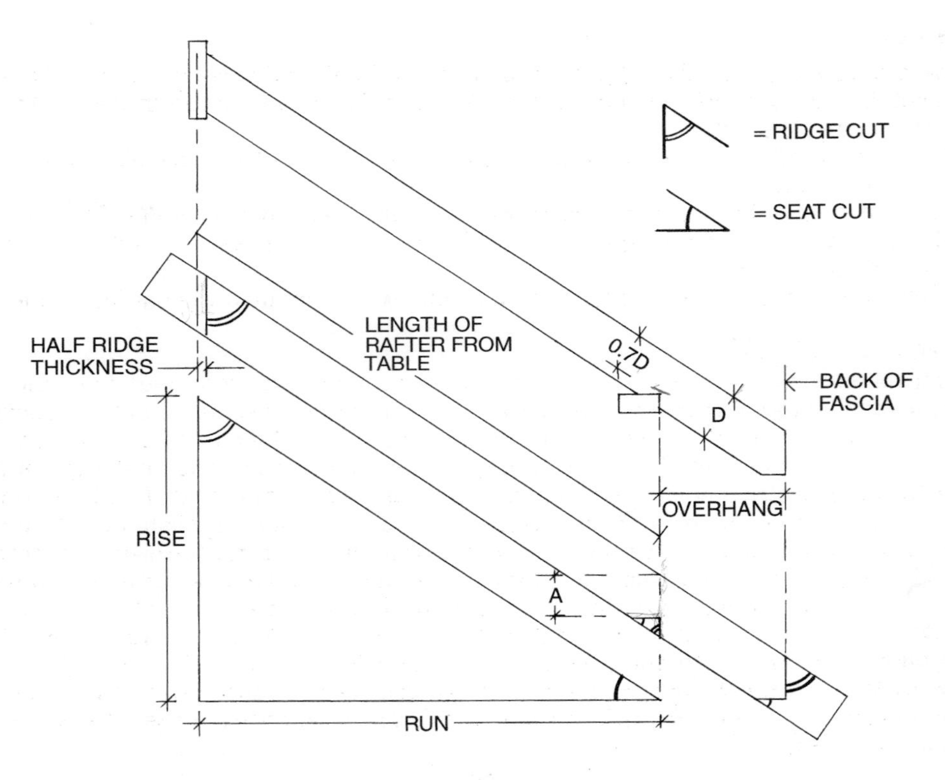

= RIDGE CUT
= SEAT CUT
LENGTH OF RAFTER FROM TABLE
HALF RIDGE THICKNESS
0.7D
D
BACK OF FASCIA
OVERHANG
RISE
A
RUN

Fig. 8.

DO NOT CUT YET!

Referring now to the tables, mark the ridge bevel on one end for the ridge cut and again at the other end for the fascia cut: in this example the angle will be 54°. The distance between those marks will be the length of the rafter calculated above.

Mark the birdsmouth as shown in Fig. 8 again using the ridge and seat bevels from the tables as indicated in the illustration.

Next mark the soffit line – this must be taken from the building design. If the soffit cut on the rafter is to be at the soffit line, the seat bevel can again be used. If the soffit line is lower than the lowest point of the rafter, then no soffit cut is required.

Now check all dimensions and angles for this first rafter which should be regarded as the master.

NOW CUT THE FIRST RAFTER

Using this first rafter check fit to the roof and use this as a pattern to mark out all of the remaining identical rafters. NB to give a true line for the fascia, it is common practice not to cut the fascia cut on all rafters at this stage. Leave the fascia cut on all roof members until the construction is complete, then using a chalked line from one end of the roof to the other, mark the fascia line on the top of the rafters. From this line, using a level, a true plumb line can be marked and the rafter cut. This traditional method involves cutting the fascia cut on the roof itself which is a time consuming task usually done using a hand saw. Trussed rafter roofs, being prefabricated, generally have the fascia cut made at the factory. If this is the case and allowing for some manufacturing tolerance on span, it will not be possible to line fascia cuts on both sides of the roof. There are two courses of action, (1) to re-mark and cut the foot of the trussed rafter again as outlined above or (2) to use packers to align the fascia onto the pre-cut feet. DO NOT be tempted to align trussed rafters to one side of the roof by aligning their rafter feet. Allowable manufacturing tolerances in a roof of the span we have been discussing could result in a variation of up to 9 mm, thus moving the ridge off the centre line, and up to a 9 mm variance between the feet of the rafters on the opposite side of the roof.

Cutting the common rafter can obviously be done by hand saw, by powered hand saw, or by using a powered compound mitre saw which can be pre-set at the ridge bevel, and then with a saw table with the length stop at an appropriate position, all cuts will be precisely the same with no further marking required.

HIP JACK RAFTERS

The length of these members will depend upon the centres at which they are to be fixed; by that we mean their spacing centre line to centre line of the thickness of the member, which should match the common rafter spacing. See Figs 1 and 9.

Continuing with the example above, the basic common rafter length was 5.2281 m (17′0″), and then assuming a jack rafter spacing of 600 mm (24″) by referring to the table, it can be seen that this length must be reduced by 742 mm 2′5$\frac{5}{8}$″.

DO NOT forget to add the over hang of the common rafter; DO NOT adjust the jack rafter for the hip until a trial fit has been taken. The jack rafter meets the hip at an angle and must therefore be cut at an angle to meet the hip both horizontally and vertically, giving what is known as a compound cut. The hip, being fixed vertically in its section, gives the same bevel cut at the top of the jack rafter as was used at the ridge and this same ridge bevel can be used. However, the edge cut can be found in the tables as the 'edge bevel', and for this an example can be seen as 39°. With these two angles the top of the hip jack can be marked, and at the lower end, the common rafter master can be used to mark the fascia cut and birdsmouth.

Cutting compound bevels by hand is a skilled task, but the powered compound mitre saw can be used to produce accurate repeatable compound cuts.

HIP RAFTER

A full hip (i.e. that which is constructed from wall plate to ridge), will have the same rise as the common rafters and the tables have been calculated on the assumption that the hip is on the mitre of a right

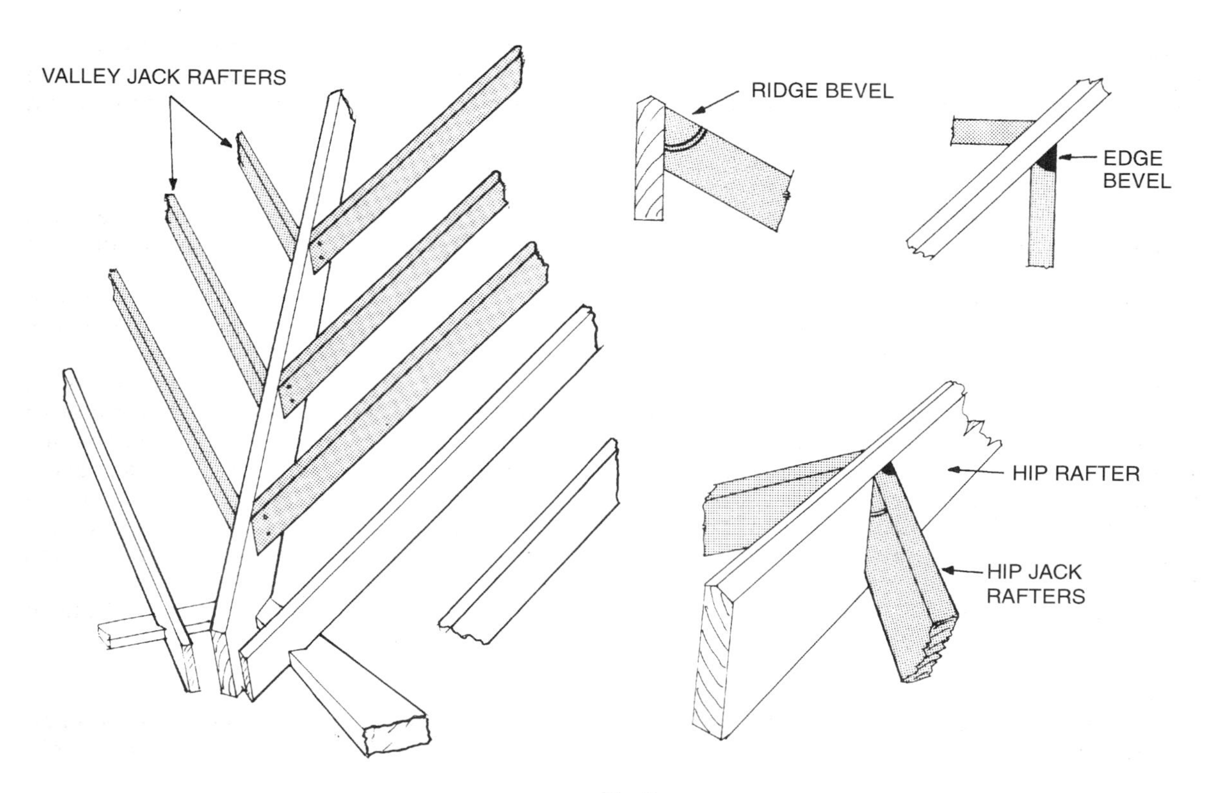
VALLEY JACK RAFTERS
RIDGE BEVEL
EDGE
BEVEL
HIP RAFTER
HIP JACK
RAFTERS

Fig. 9.

angled corner and is therefore at right angles to the common rafters. This allows the same run as the common rafter to be used to save further calculation.

Continuing with the example above then, the run of the hip would be 4.23 m (13′10½″), and by referring to the tables for the length of hip, the calculation will result in a hip length of 6.7257 m (22′0⅝″).

The seat and ridge bevels can be taken directly from the tables in this case 27° and 63° respectively. Care must be taken when setting out the birdsmouth to ensure that the depth of rafter (A) in the illustration Fig. 10, equals that of the common rafter illustrated in Fig. 8.

The mitre at the top of the hips where they meet the ridge does need the special setting of a bevel. The marking gauge is set to ½ the thickness of the hip and marked on the end of both faces after the plumb cut is made. See Fig. 10.

Backing of Hips

In a good job, the hips are backed – that is to say a chamfer is planed both ways from the centre line on the top of the hip so that the two surfaces are in line with the planes of the roof on adjacent sides. This gives a good seating for the battens. After cutting the hip to the plumb line the same plumb bevel for marking the profile of the backing chamfers on each side of the hip may be used. See Fig. 10.

Dimension B, the length of the plumb cut of the jack rafters is measured on the top end of the hip down from the backing levels leaving a remainder C. If C is measured along the side bevel of purlin it gives the position of the projection under the hip. See Fig. 11.

VALLEY JACK RAFTERS

The tables are based on a construction which assumes a valley rafter similar to the hip rafter, see Fig. 9, NOT that illustrated in Fig. 4 which is a more modern construction and one which works with a trussed rafter roof if a prefabricated valley set of frames is not provided. Returning then to the traditional cut valley, this is essentially a hip in reverse. The valley jacks decrease in length as they progress up the roof, and again would be based at similar centres to the common rafters. The same bevels as for the hip jack

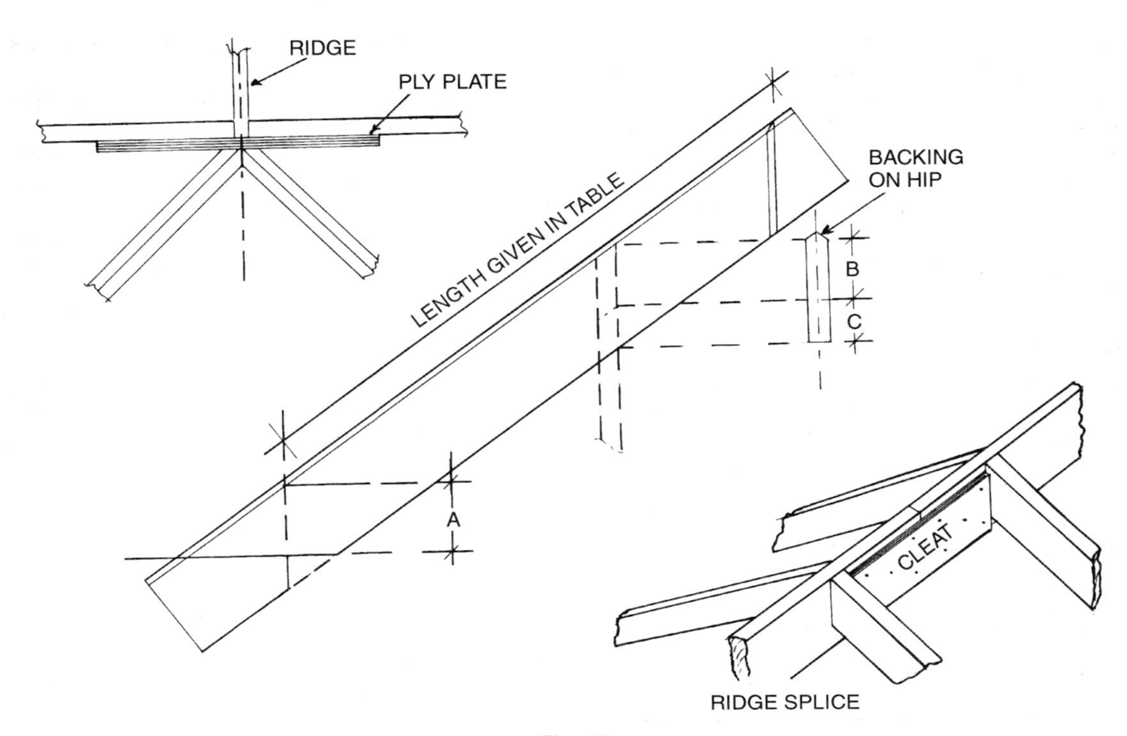
RIDGE
PLY PLATE
LENGTH GIVEN IN TABLE
BACKING
ON HIP
B
C
A
CLEAT
RIDGE SPLICE

Fig. 10.

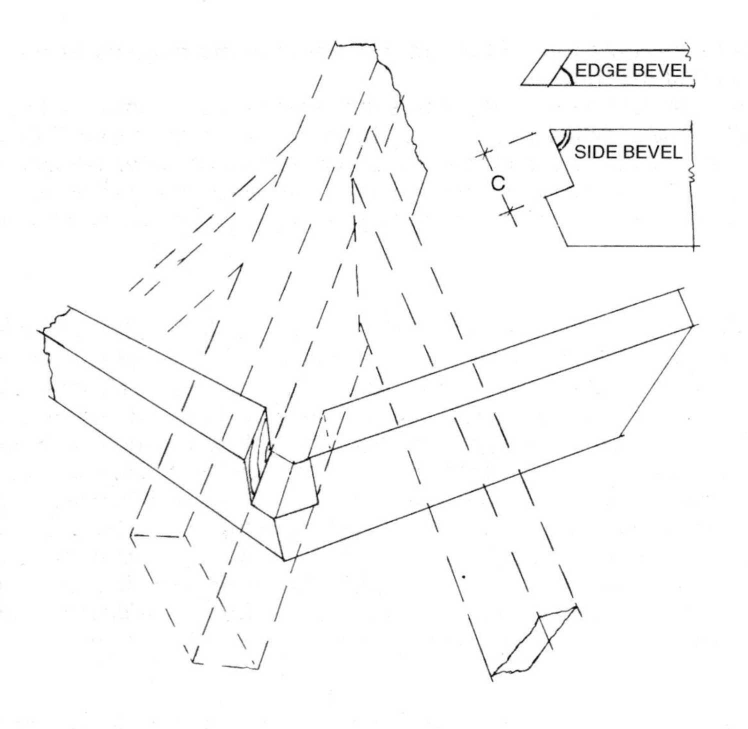

Fig. 11.

rafter can be used but this time on the foot of the rafter rather than the ridge as before. The common rafter ridge bevel can be used at the top.

Now referring to Fig. 4, the top cut on the valley jack is the same as the common rafter, but the foot of the valley rests on a flat valley board nailed on the top of the rafters of the main roof. This construction is suitable for all valleys except attic construction. The cut at the bottom of the valley jack is the same as the seat cut for the common rafter with an edge bevel equal to the pitch of the main roof. This may not be the same as the pitch of the roof on which the valley jack has to be fitted and should therefore be checked.

THE RIDGE

This roof member is usually relatively thin and can be no more than 25–38 mm (1″–1½″). It takes little load from the roof as pairs of common rafters or valley rafters are placed directly opposite one another across the ridge, it does act as a tie from gable to gable or hip to hip. In a roof with a gable at both ends, the length of the ridge equals the length of the wall plate and is usually built into the gables at both ends. In a hip roof with a hip at both ends, the ridge length is normally the length of the building (internally) less the width (internally). If a short packing piece is used at the end of the ridge in a hip construction to give good hip rafter supports, then this thickness must be deducted from the overall length of the ridge for each hip. The packing piece should be at least as deep as the hip rafter ridge cut, and may therefore be too deep for readily available softwood. The use of 19 mm (¾″) exterior grade plywood is recommended. The ply can also be used to couple ridge pieces together in their length because even the longest lengths of timber may be too short for the overall length of the ridge itself. See Fig. 10. These 'cleats' should be located between rafters to avoid having to cut back the rafter by the thickness of the cleat.

PURLINS

As illustrated under the definition for purlin earlier in this book, Fig. 1, the purlin can be fitted either at right angles to the underside of the rafter or vertically. The tables given in this book show edge and side bevels for purlins set at right angles to the rafters. See Fig. 11. This is the traditional form of construction and may have to be followed on extension work to older buildings.

The purlin fitted vertically is preferred from a structural viewpoint because it acts as a true beam carrying the rafters. The rafters can be better fitted by birdsmouth to the purlin. The edge bevel on this type of purlin is that formed by the hip on plan which on a right angled hip is 45°. The side bevel is 90°, simply a square cut.

4 METRIC CALCULATION TABLES

RISE OF COMMON RAFTER 0.087 m **PER METRE OF RUN** **PITCH** 5°

BEVELS:	COMMON RAFTER	– SEAT	5
	" "	– RIDGE	85
	HIP OR VALLEY	– SEAT	
	" " "	– RIDGE	
	JACK RAFTER	– EDGE	
	PURLIN	– EDGE	
	"	– SIDE	

JACK RAFTERS	333 mm	**CENTRES**	**DECREASE**	(in mm to 999 and thereafter in m)
	400 "	"	"	
	500 "	"	"	
	600 "	"	"	

Run of Rafter	0.1	0.2	0.3	0.4	0.5	0.6	0.7	0.8	0.9	1.0
Length of Rafter	0.1	0.201	0.301	0.401	0.502	0.602	0.703	0.803	0.903	1.004
Length of Hip										

RISE OF COMMON RAFTER 0.105 m **PER METRE OF RUN** **PITCH** 6°

BEVELS:	COMMON RAFTER	– SEAT	6
	" "	– RIDGE	84
	HIP OR VALLEY	– SEAT	
	" " "	– RIDGE	
	JACK RAFTER	– EDGE	
	PURLIN	– EDGE	
	"	– SIDE	

JACK RAFTERS 333 mm **CENTRES DECREASE** (in mm to 999 and thereafter in m)
400 " " "
500 " " "
600 " " "

Run of Rafter	0.1	0.2	0.3	0.4	0.5	0.6	0.7	0.8	0.9	1.0
Length of Rafter	0.101	0.201	0.302	0.402	0.503	0.603	0.704	0.804	0.905	1.006
Length of Hip										

RISE OF COMMON RAFTER 0.123 m PER METRE OF RUN   PITCH 7°

BEVELS:	COMMON RAFTER	– SEAT	7
	" "	– RIDGE	83
	HIP OR VALLEY	– SEAT	
	" " "	– RIDGE	
	JACK RAFTER	– EDGE	
	PURLIN	– EDGE	
	"	– SIDE	

JACK RAFTERS	333 mm	CENTRES	DECREASE	(in mm to 999 and thereafter in m)
	400 "	"	"	
	500 "	"	"	
	600 "	"	"	

Run of Rafter	0.1	0.2	0.3	0.4	0.5	0.6	0.7	0.8	0.9	1.0
Length of Rafter	0.101	0.202	0.302	0.403	0.504	0.605	0.705	0.806	0.907	1.008
Length of Hip										

RISE OF COMMON RAFTER 0.141 m **PER METRE OF RUN** **PITCH** 8°

BEVELS:	COMMON RAFTER	– SEAT	8
	" "	– RIDGE	82
	HIP OR VALLEY	– SEAT	
	" " "	– RIDGE	
	JACK RAFTER	– EDGE	
	PURLIN	– EDGE	
	"	– SIDE	

JACK RAFTERS 333 mm **CENTRES DECREASE** (in mm to 999 and thereafter in m)
400 " " "
500 " " "
600 " " "

Run of Rafter	0.1	0.2	0.3	0.4	0.5	0.6	0.7	0.8	0.9	1.0
Length of Rafter	0.101	0.202	0.303	0.404	0.505	0.606	0.707	0.808	0.909	1.01
Length of Hip										

RISE OF COMMON RAFTER 0.158 m **PER METRE OF RUN** **PITCH** 9°

BEVELS:	COMMON RAFTER	– SEAT	9
	" "	– RIDGE	81
	HIP OR VALLEY	– SEAT	
	" " "	– RIDGE	
	JACK RAFTER	– EDGE	
	PURLIN	– EDGE	
	"	– SIDE	

JACK RAFTERS	333 mm **CENTRES DECREASE**	(in mm to 999 and thereafter in m)
	400 " " "	
	500 " " "	
	600 " " "	

Run of Rafter	0.1	0.2	0.3	0.4	0.5	0.6	0.7	0.8	0.9	1.0
Length of Rafter	0.101	0.203	0.304	0.405	0.506	0.608	0.709	0.81	0.911	1.013
Length of Hip										

RISE OF COMMON RAFTER 0.176 m **PER METRE OF RUN** **PITCH** 10°

BEVELS:	COMMON RAFTER	– SEAT	10
	" "	– RIDGE	80
	HIP OR VALLEY	– SEAT	
	" " "	– RIDGE	
	JACK RAFTER	– EDGE	
	PURLIN	– EDGE	
	"	– SIDE	

JACK RAFTERS	333 mm **CENTRES DECREASE**	(in mm to 999 and thereafter in m)
	400 " " "	
	500 " " "	
	600 " " "	

Run of Rafter	0.1	0.2	0.3	0.4	0.5	0.6	0.7	0.8	0.9	1.0
Length of Rafter	0.102	0.203	0.305	0.406	0.508	0.609	0.711	0.812	0.914	1.015
Length of Hip										

RISE OF COMMON RAFTER 0.194 m **PER METRE OF RUN** **PITCH** 11°

BEVELS:	COMMON RAFTER	– SEAT	11
	" "	– RIDGE	79
	HIP OR VALLEY	– SEAT	
	" " "	– RIDGE	
	JACK RAFTER	– EDGE	
	PURLIN	– EDGE	
	"	– SIDE	

JACK RAFTERS	333 mm	**CENTRES**	**DECREASE**
	400 "	"	"
	500 "	"	"
	600 "	"	"

(in mm to 999 and thereafter in m)

Run of Rafter	0.1	0.2	0.3	0.4	0.5	0.6	0.7	0.8	0.9	1.0
Length of Rafter	0.102	0.204	0.306	0.407	0.509	0.611	0.713	0.815	0.917	1.019
Length of Hip										

RISE OF COMMON RAFTER 0.213 m **PER METRE OF RUN** **PITCH** 12°

BEVELS:	COMMON RAFTER	– SEAT	12
	" "	– RIDGE	78
	HIP OR VALLEY	– SEAT	
	" " "	– RIDGE	
	JACK RAFTER	– EDGE	
	PURLIN	– EDGE	
	"	– SIDE	

JACK RAFTERS	333 mm	**CENTRES DECREASE**	(in mm to 999 and thereafter in m)
	400 "	" "	
	500 "	" "	
	600 "	" "	

Run of Rafter	0.1	0.2	0.3	0.4	0.5	0.6	0.7	0.8	0.9	1.0
Length of Rafter	0.102	0.204	0.307	0.409	0.511	0.613	0.716	0.818	0.92	1.022
Length of Hip										

RISE OF COMMON RAFTER 0.231 m **PER METRE OF RUN** **PITCH** 13°

BEVELS:	COMMON RAFTER	– SEAT	13
	" "	– RIDGE	77
	HIP OR VALLEY	– SEAT	
	" " "	– RIDGE	
	JACK RAFTER	– EDGE	
	PURLIN	– EDGE	
	"	– SIDE	

JACK RAFTERS	333 mm **CENTRES DECREASE**	(in mm to 999 and thereafter in m)
	400 " " "	
	500 " " "	
	600 " " "	

Run of Rafter	0.1	0.2	0.3	0.4	0.5	0.6	0.7	0.8	0.9	1.0
Length of Rafter	0.103	0.205	0.308	0.411	0.513	0.616	0.718	0.821	0.924	1.026
Length of Hip										

RISE OF COMMON RAFTER 0.249 m PER METRE OF RUN — PITCH 14°

BEVELS:		
	COMMON RAFTER	– SEAT 14
	" "	– RIDGE 76
	HIP OR VALLEY	– SEAT
	" " "	– RIDGE
	JACK RAFTER	– EDGE
	PURLIN	– EDGE
	"	– SIDE

JACK RAFTERS			
333 mm	CENTRES	DECREASE	(in mm to 999 and thereafter in m)
400 "	"	"	
500 "	"	"	
600 "	"	"	

Run of Rafter	0.1	0.2	0.3	0.4	0.5	0.6	0.7	0.8	0.9	1.0
Length of Rafter	0.103	0.206	0.309	0.412	0.515	0.618	0.721	0.824	0.928	1.031
Length of Hip										

RISE OF COMMON RAFTER 0.268 m **PER METRE OF RUN** **PITCH** 15°

BEVELS:	COMMON RAFTER	– SEAT	15
	" "	– RIDGE	75
	HIP OR VALLEY	– SEAT	
	" " "	– RIDGE	
	JACK RAFTER	– EDGE	
	PURLIN	– EDGE	
	"	– SIDE	

JACK RAFTERS	333 mm **CENTRES DECREASE**	(in mm to 999 and thereafter in m)
	400 " " "	
	500 " " "	
	600 " " "	

Run of Rafter	0.1	0.2	0.3	0.4	0.5	0.6	0.7	0.8	0.9	1.0
Length of Rafter	0.104	0.207	0.311	0.414	0.518	0.621	0.725	0.828	0.932	1.035
Length of Hip										

RISE OF COMMON RAFTER 0.287 m **PER METRE OF RUN** **PITCH** 16°
(Grecian pitch)

BEVELS:	COMMON RAFTER	–	SEAT	16
	" "	–	RIDGE	74
	HIP OR VALLEY	–	SEAT	11.5
	" " "	–	RIDGE	78.5
	JACK RAFTER	–	EDGE	44
	PURLIN	–	EDGE	46
	"	–	SIDE	74.5

JACK RAFTERS	333 mm	**CENTRES**	**DECREASE**	346	(in mm to 999 and thereafter in m)
	400 "	"	"	416	
	500 "	"	"	520	
	600 "	"	"	624	

Run of Rafter	0.1	0.2	0.3	0.4	0.5	0.6	0.7	0.8	0.9	1.0
Length of Rafter	0.104	0.208	0.312	0.416	0.52	0.624	0.728	0.832	0.936	1.04
Length of Hip	0.144	0.289	0.433	0.577	0.722	0.866	1.01	1.154	1.299	1.443

RISE OF COMMON RAFTER 0.306 m **PER METRE OF RUN** **PITCH** 17°

BEVELS:			
	COMMON RAFTER	– SEAT	17
	" "	– RIDGE	73
	HIP OR VALLEY	– SEAT	12
	" " "	– RIDGE	78
	JACK RAFTER	– EDGE	43.5
	PURLIN	– EDGE	46.5
	"	– SIDE	73.5

JACK RAFTERS		**CENTRES DECREASE**		
	333 mm		348	(in mm to 999 and thereafter in m)
	400 "	" "	418	
	500 "	" "	522	
	600 "	" "	627	

Run of Rafter	0.1	0.2	0.3	0.4	0.5	0.6	0.7	0.8	0.9	1.0
Length of Rafter	0.105	0.209	0.314	0.418	0.523	0.627	0.732	0.837	0.941	1.046
Length of Hip	0.145	0.289	0.434	0.579	0.723	0.868	1.013	1.158	1.302	1.447

RISE OF COMMON RAFTER 0.315 m PER METRE OF RUN   PITCH $17\frac{1}{2}°$

BEVELS:	COMMON RAFTER	– SEAT	17.5
	" "	– RIDGE	72.5
	HIP OR VALLEY	– SEAT	12.5
	" " "	– RIDGE	77.5
	JACK RAFTER	– EDGE	43.5
	PURLIN	– EDGE	46.5
	"	– SIDE	73.5

JACK RAFTERS	333 mm	**CENTRES DECREASE**	349	(in mm to 999 and thereafter in m)
	400 "	"	420	
	500 "	"	524	
	600 "	"	629	

Run of Rafter	0.1	0.2	0.3	0.4	0.5	0.6	0.7	0.8	0.9	1.0
Length of Rafter	0.105	0.210	0.315	0.419	0.524	0.629	0.734	0.839	0.944	1.049
Length of Hip	0.145	0.29	0.435	0.579	0.724	0.829	1.014	1.159	1.304	1.449

RISE OF COMMON RAFTER 0.325 m **PER METRE OF RUN** **PITCH** 18°

BEVELS:			
	COMMON RAFTER	– SEAT	18
	" "	– RIDGE	72
	HIP OR VALLEY	– SEAT	13
	" " "	– RIDGE	77
	JACK RAFTER	– EDGE	43.5
	PURLIN	– EDGE	46.5
	"	– SIDE	73

JACK RAFTERS	333 mm	CENTRES DECREASE	350	(in mm to 999 and thereafter in m)
	400 "	" "	421	
	500 "	" "	526	
	600 "	" "	631	

Run of Rafter	0.1	0.2	0.3	0.4	0.5	0.6	0.7	0.8	0.9	1.0
Length of Rafter	0.105	0.21	0.315	0.421	0.526	0.631	0.736	0.841	0.946	1.051
Length of Hip	0.145	0.29	0.435	0.58	0.726	0.871	1.016	1.161	1.306	1.451

RISE OF COMMON RAFTER 0.344 m PER METRE OF RUN   PITCH 19°

BEVELS:			
	COMMON RAFTER	– SEAT	19
	" "	– RIDGE	71
	HIP OR VALLEY	– SEAT	13.5
	" " "	– RIDGE	76.5
	JACK RAFTER	– EDGE	43.5
	PURLIN	– EDGE	46.5
	"	– SIDE	72

JACK RAFTERS				
333 mm	CENTRES	DECREASE	352	(in mm to 999 and thereafter in m)
400 "	"	"	423	
500 "	"	"	529	
600 "	"	"	635	

Run of Rafter	0.1	0.2	0.3	0.4	0.5	0.6	0.7	0.8	0.9	1.0
Length of Rafter	0.106	0.212	0.317	0.423	0.529	0.635	0.74	0.846	0.952	1.058
Length of Hip	0.146	0.291	0.437	0.582	0.728	0.873	1.019	1.164	1.31	1.456

RISE OF COMMON RAFTER 0.364 m **PER METRE OF RUN** **PITCH** 20°

BEVELS:	COMMON RAFTER	– SEAT	20
	" "	– RIDGE	70
	HIP OR VALLEY	– SEAT	14.5
	" " "	– RIDGE	75.5
	JACK RAFTER	– EDGE	43
	PURLIN	– EDGE	47
	"	– SIDE	71

JACK RAFTERS	333 mm	**CENTRES**	**DECREASE**	354	(in mm to 999 and thereafter in m)
	400 "	"	"	426	
	500 "	"	"	532	
	600 "	"	"	638	

Run of Rafter	0.1	0.2	0.3	0.4	0.5	0.6	0.7	0.8	0.9	1.0
Length of Rafter	0.106	0.213	0.319	0.426	0.532	0.639	0.745	0.851	0.958	1.064
Length of Hip	0.146	0.292	0.438	0.584	0.73	0.876	1.022	1.168	1.314	1.46

RISE OF COMMON RAFTER 0.384 m PER METRE OF RUN   PITCH 21°

BEVELS:	COMMON RAFTER	– SEAT	21
	〃 〃	– RIDGE	69
	HIP OR VALLEY	– SEAT	15
	〃 〃 〃	– RIDGE	75
	JACK RAFTER	– EDGE	43
	PURLIN	– EDGE	47
	〃	– SIDE	70.5

JACK RAFTERS	333 mm	CENTRES DECREASE	357	(in mm to 999 and thereafter in m)
	400 〃	〃 〃	428	
	500 〃	〃 〃	536	
	600 〃	〃 〃	643	

Run of Rafter	0.1	0.2	0.3	0.4	0.5	0.6	0.7	0.8	0.9	1.0
Length of Rafter	0.107	0.214	0.321	0.428	0.536	0.643	0.75	0.857	0.964	1.071
Length of Hip	0.147	0.293	0.44	0.586	0.733	0.879	1.026	1.172	1.319	1.465

RISE OF COMMON RAFTER 0.404 m **PER METRE OF RUN** **PITCH** 22°

BEVELS:	COMMON RAFTER	– SEAT	22
	" "	– RIDGE	68
	HIP OR VALLEY	– SEAT	16
	" " "	– RIDGE	74
	JACK RAFTER	– EDGE	43
	PURLIN	– EDGE	47
	"	– SIDE	69.5

JACK RAFTERS				
333 mm	**CENTRES**	**DECREASE**	359	(in mm to 999 and thereafter in m)
400 "	"	"	432	
500 "	"	"	540	
600 "	"	"	647	

Run of Rafter	0.1	0.2	0.3	0.4	0.5	0.6	0.7	0.8	0.9	1.0
Length of Rafter	0.108	0.216	0.324	0.431	0.539	0.647	0.755	0.863	0.971	1.079
Length of Hip	0.147	0.294	0.441	0.588	0.736	0.883	1.03	1.177	1.324	1.471

RISE OF COMMON RAFTER 0.414 m PER METRE OF RUN PITCH $22\frac{1}{2}°$

BEVELS:			
	COMMON RAFTER	– SEAT	22.5
	" "	– RIDGE	67.5
	HIP OR VALLEY	– SEAT	16.25
	" " "	– RIDGE	73.75
	JACK RAFTER	– EDGE	42.75
	PURLIN	– EDGE	47.5
	"	– SIDE	69.0

JACK RAFTERS			
333 mm	CENTRES DECREASE	361	(in mm to 999 and thereafter in m)
400 "	" "	433	
500 "	" "	542	
600 "	" "	650	

Run of Rafter	0.1	0.2	0.3	0.4	0.5	0.6	0.7	0.8	0.9	1.0
Length of Rafter	0.108	0.216	0.325	0.433	0.541	0.649	0.758	0.866	0.974	1.082
Length of Hip	0.147	0.294	0.442	0.589	0.737	0.884	1.032	1.179	1.326	1.473

RISE OF COMMON RAFTER 0.424 m **PER METRE OF RUN** **PITCH** 23°

BEVELS:	COMMON RAFTER	– SEAT	23
	" "	– RIDGE	67
	HIP OR VALLEY	– SEAT	16.5
	" " "	– RIDGE	73.5
	JACK RAFTER	– EDGE	42.5
	PURLIN	– EDGE	47.5
	"	– SIDE	68.5

JACK RAFTERS	333 mm	**CENTRES DECREASE**	362	(in mm to 999 and
	400 "	" "	434	thereafter in m)
	500 "	" "	543	
	600 "	" "	652	

Run of Rafter	0.1	0.2	0.3	0.4	0.5	0.6	0.7	0.8	0.9	1.0
Length of Rafter	0.109	0.217	0.326	0.435	0.543	0.652	0.76	0.869	0.978	1.086
Length of Hip	0.148	0.295	0.443	0.591	0.739	0.886	1.034	1.181	1.329	1.477

RISE OF COMMON RAFTER 0.445 m **PER METRE OF RUN** **PITCH** 24°
(Roman pitch)

BEVELS:	COMMON RAFTER	– SEAT	24
	〃 〃	– RIDGE	66
	HIP OR VALLEY	– SEAT	17.5
	〃 〃 〃	– RIDGE	72.5
	JACK RAFTER	– EDGE	42.5
	PURLIN	– EDGE	47.5
	〃	– SIDE	68

JACK RAFTERS	333 mm	**CENTRES**	**DECREASE**	365	(in mm to 999 and thereafter in m)
	400 〃	〃	〃	438	
	500 〃	〃	〃	548	
	600 〃	〃	〃	657	

Run of Rafter	0.1	0.2	0.3	0.4	0.5	0.6	0.7	0.8	0.9	1.0
Length of Rafter	0.109	0.219	0.328	0.438	0.547	0.657	0.766	0.876	0.985	1.095
Length of Hip	0.148	0.297	0.445	0.593	0.741	0.89	1.038	1.186	1.334	1.483

RISE OF COMMON RAFTER 0.466 m **PER METRE OF RUN** **PITCH** 25°

BEVELS:			
	COMMON RAFTER	– SEAT	25
	" "	– RIDGE	65
	HIP OR VALLEY	– SEAT	18
	" " "	– RIDGE	72
	JACK RAFTER	– EDGE	42
	PURLIN	– EDGE	48
	"	– SIDE	67

JACK RAFTERS			
333 mm	**CENTRES DECREASE**	367	(in mm to 999 and thereafter in m)
400 "	" "	441	
500 "	" "	552	
600 "	" "	662	

Run of Rafter	0.1	0.2	0.3	0.4	0.5	0.6	0.7	0.8	0.9	1.0
Length of Rafter	0.11	0.221	0.331	0.441	0.552	0.662	0.772	0.883	0.993	1.103
Length of Hip	0.149	0.298	0.447	0.596	0.745	0.893	1.042	1.191	1.34	1.489

RISE OF COMMON RAFTER 0.5 m **PER METRE OF RUN** **PITCH** 26° 34′

(Quarter pitch)

BEVELS:	COMMON RAFTER	–	SEAT	26.5
	〃 〃	–	RIDGE	63.5
	HIP OR VALLEY	–	SEAT	19.5
	〃 〃 〃	–	RIDGE	70.5
	JACK RAFTER	–	EDGE	42
	PURLIN	–	EDGE	48
	〃	–	SIDE	66

JACK RAFTERS	333 mm	**CENTRES DECREASE**	372	(in mm to 999 and thereafter in m)
	400 〃	〃 〃	447	
	500 〃	〃 〃	559	
	600 〃	〃 〃	671	

Run of Rafter	0.1	0.2	0.3	0.4	0.5	0.6	0.7	0.8	0.9	1.0
Length of Rafter	0.112	0.224	0.335	0.447	0.559	0.671	0.783	0.894	1.006	1.118
Length of Hip	0.15	0.3	0.45	0.6	0.75	0.9	1.05	1.2	1.35	1.5

RISE OF COMMON RAFTER 0.521 m **PER METRE OF RUN** **PITCH** $27\frac{1}{2}^{\circ}$

BEVELS:	COMMON RAFTER	– SEAT	27.5
	" "	– RIDGE	62.5
	HIP OR VALLEY	– SEAT	20
	" " "	– RIDGE	70
	JACK RAFTER	– EDGE	41.75
	PURLIN	– EDGE	48.5
	"	– SIDE	65

JACK RAFTERS	333 mm	**CENTRES**	**DECREASE**	375	(in mm to 999 and thereafter in m)
	400 "	"	"	451	
	500 "	"	"	563	
	600 "	"	"	677	

Run of Rafter	0.1	0.2	0.3	0.4	0.5	0.6	0.7	0.8	0.9	1.0
Length of Rafter	0.113	0.225	0.338	0.451	0.564	0.676	0.789	0.902	1.015	1.127
Length of Hip	0.151	0.301	0.452	0.603	0.754	0.904	1.054	1.206	1.365	1.507

RISE OF COMMON RAFTER 0.532 m **PER METRE OF RUN** **PITCH** 28°

BEVELS:	COMMON RAFTER	– SEAT	28
	" "	– RIDGE	62
	HIP OR VALLEY	– SEAT	20.5
	" " "	– RIDGE	69.5
	JACK RAFTER	– EDGE	41.5
	PURLIN	– EDGE	48.5
	"	– SIDE	65

JACK RAFTERS	333 mm	**CENTRES DECREASE**	377	(in mm to 999 and thereafter in m)
	400 "	" "	453	
	500 "	" "	566	
	600 "	" "	680	

Run of Rafter	0.1	0.2	0.3	0.4	0.5	0.6	0.7	0.8	0.9	1.0
Length of Rafter	0.113	0.227	0.34	0.453	0.566	0.68	0.793	0.906	1.019	1.133
Length of Hip	0.151	0.302	0.453	0.603	0.754	0.905	1.056	1.207	1.358	1.511

RISE OF COMMON RAFTER 0.544 m **PER METRE OF RUN** **PITCH** 29°

BEVELS:	COMMON RAFTER	– SEAT	29
	" "	– RIDGE	61
	HIP OR VALLEY	– SEAT	21.5
	" " "	– RIDGE	68.5
	JACK RAFTER	– EDGE	41
	PURLIN	– EDGE	49
	"	– SIDE	64

JACK RAFTERS			
333 mm	**CENTRES DECREASE**	381	(in mm to 999 and thereafter in m)
400 "	" "	457	
500 "	" "	572	
600 "	" "	686	

Run of Rafter	0.1	0.2	0.3	0.4	0.5	0.6	0.7	0.8	0.9	1.0
Length of Rafter	0.114	0.229	0.343	0.457	0.572	0.686	0.8	0.914	1.029	1.143
Length of Hip	0.152	0.304	0.456	0.608	0.759	0.912	1.063	1.215	1.367	1.519

RISE OF COMMON RAFTER 0.577 m **PER METRE OF RUN** **PITCH** 30°

BEVELS:	COMMON RAFTER	– SEAT	30
	" "	– RIDGE	60
	HIP OR VALLEY	– SEAT	22
	" " "	– RIDGE	68
	JACK RAFTER	– EDGE	41
	PURLIN	– EDGE	49
	"	– SIDE	63.5

JACK RAFTERS	333 mm	**CENTRES DECREASE**	385	(in mm to 999 and thereafter in m)
	400 "	" "	462	
	500 "	" "	577	
	600 "	" "	693	

Run of Rafter	0.1	0.2	0.3	0.4	0.5	0.6	0.7	0.8	0.9	1.0
Length of Rafter	0.116	0.231	0.346	0.462	0.577	0.693	0.808	0.924	1.039	1.155
Length of Hip	0.153	0.306	0.458	0.611	0.764	0.917	1.069	1.222	1.375	1.528

RISE OF COMMON RAFTER 0.601 m **PER METRE OF RUN** **PITCH** 31°

BEVELS:			
	COMMON RAFTER	– SEAT	31
	" "	– RIDGE	59
	HIP OR VALLEY	– SEAT	23
	" " "	– RIDGE	67
	JACK RAFTER	– EDGE	40.5
	PURLIN	– EDGE	49.5
	"	– SIDE	62.5

JACK RAFTERS			
333 mm	**CENTRES DECREASE**	389	(in mm to 999 and thereafter in m)
400 "	" "	467	
500 "	" "	584	
600 "	" "	700	

Run of Rafter	0.1	0.2	0.3	0.4	0.5	0.6	0.7	0.8	0.9	1.0
Length of Rafter	0.117	0.233	0.35	0.467	0.583	0.7	0.817	0.933	1.05	1.167
Length of Hip	0.154	0.307	0.461	0.615	0.768	0.922	1.076	1.229	1.383	1.537

RISE OF COMMON RAFTER 0.625 m **PER METRE OF RUN** **PITCH** 32°

BEVELS:	COMMON RAFTER	– SEAT	32
	" "	– RIDGE	58
	HIP OR VALLEY	– SEAT	24
	" " "	– RIDGE	66
	JACK RAFTER	– EDGE	40.5
	PURLIN	– EDGE	49.5
	"	– SIDE	62

JACK RAFTERS	333 mm	**CENTRES DECREASE**	393	(in mm to 999 and thereafter in m)
	400 "	" "	472	
	500 "	" "	590	
	600 "	" "	707	

Run of Rafter	0.1	0.2	0.3	0.4	0.5	0.6	0.7	0.8	0.9	1.0
Length of Rafter	0.118	0.239	0.354	0.472	0.59	0.708	0.825	0.943	1.061	1.179
Length of Hip	0.155	0.309	0.464	0.618	0.773	0.928	1.082	1.237	1.391	1.546

RISE OF COMMON RAFTER 0.637 m **PER METRE OF RUN** **PITCH** $32\frac{1}{2}°$

BEVELS:	COMMON RAFTER	– SEAT	32.5
	" "	– RIDGE	57.5
	HIP OR VALLEY	– SEAT	24.25
	" " "	– RIDGE	65.75
	JACK RAFTER	– EDGE	40.25
	PURLIN	– EDGE	49.75
	"	– SIDE	61.5

JACK RAFTERS	333 mm	**CENTRES**	**DECREASE**	395	(in mm to 999 and thereafter in m)
	400 "	"	"	475	
	500 "	"	"	593	
	600 "	"	"	711	

Run of Rafter	0.1	0.2	0.3	0.4	0.5	0.6	0.7	0.8	0.9	1.0
Length of Rafter	0.119	0.237	0.356	0.474	0.593	0.711	0.830	0.949	1.067	1.186
Length of Hip	0.155	0.310	0.466	0.620	0.776	0.930	1.086	1.241	1.396	1.551

RISE OF COMMON RAFTER 0.649 m PER METRE OF RUN **PITCH** 33°

BEVELS:	COMMON RAFTER	– SEAT	33
	" "	– RIDGE	57
	HIP OR VALLEY	– SEAT	24.5
	" " "	– RIDGE	65.5
	JACK RAFTER	– EDGE	40
	PURLIN	– EDGE	50
	"	– SIDE	61.5

JACK RAFTERS	333 mm	**CENTRES DECREASE**	397	(in mm to 999 and thereafter in m)
	400 "	" "	477	
	500 "	" "	596	
	600 "	" "	715	

Run of Rafter	0.1	0.2	0.3	0.4	0.5	0.6	0.7	0.8	0.9	1.0
Length of Rafter	0.119	0.238	0.358	0.48	0.596	0.715	0.835	0.954	1.073	1.192
Length of Hip	0.156	0.311	0.467	0.623	0.778	0.934	1.089	1.245	1.401	1.556

RISE OF COMMON RAFTER 0.666 m **PER METRE OF RUN** **PITCH** 33° 40′
(Third pitch)

BEVELS:	COMMON RAFTER	– SEAT	33.5
	" "	– RIDGE	56.5
	HIP OR VALLEY	– SEAT	25
	" " "	– RIDGE	65
	JACK RAFTER	– EDGE	40
	PURLIN	– EDGE	50
	"	– SIDE	61

JACK RAFTERS			
333 mm	**CENTRES DECREASE**	397	(in mm to 999 and thereafter in m)
400 "	" "	481	
500 "	" "	601	
600 "	" "	721	

Run of Rafter	0.1	0.2	0.3	0.4	0.5	0.6	0.7	0.8	0.9	1.0
Length of Rafter	0.12	0.24	0.361	0.481	0.601	0.721	0.841	0.961	1.082	1.202
Length of Hip	0.157	0.313	0.47	0.626	0.782	0.938	1.094	1.251	1.408	1.563

RISE OF COMMON RAFTER 0.7 m PER METRE OF RUN — PITCH 35°

BEVELS:			
	COMMON RAFTER	– SEAT	35
	" "	– RIDGE	55
	HIP OR VALLEY	– SEAT	26.5
	" " "	– RIDGE	63.5
	JACK RAFTER	– EDGE	39.5
	PURLIN	– EDGE	50.5
	"	– SIDE	60

JACK RAFTERS			
333 mm	CENTRES DECREASE	407	(in mm to 999 and thereafter in m)
400 "	" "	488	
500 "	" "	611	
600 "	" "	733	

Run of Rafter	0.1	0.2	0.3	0.4	0.5	0.6	0.7	0.8	0.9	1.0
Length of Rafter	0.122	0.244	0.366	0.488	0.61	0.732	0.855	0.977	1.099	1.221
Length of Hip	0.158	0.316	0.473	0.631	0.789	0.947	1.105	1.262	1.42	1.578

RISE OF COMMON RAFTER 0.727 m **PER METRE OF RUN** **PITCH** 36°

BEVELS:			
	COMMON RAFTER	– SEAT	36
	" "	– RIDGE	54
	HIP OR VALLEY	– SEAT	27
	" " "	– RIDGE	63
	JACK RAFTER	– EDGE	39
	PURLIN	– EDGE	51
	"	– SIDE	59.5

JACK RAFTERS				
333 mm	**CENTRES DECREASE**	412	(in mm to 999 and thereafter in m)
400 "	" "	494	
500 "	" "	618	
600 "	" "	742	

Run of Rafter	0.1	0.2	0.3	0.4	0.5	0.6	0.7	0.8	0.9	1.0
Length of Rafter	0.124	0.247	0.371	0.494	0.618	0.742	0.865	0.989	1.112	1.236
Length of Hip	0.159	0.318	0.477	0.636	0.795	0.954	1.113	1.272	1.431	1.59

RISE OF COMMON RAFTER 0.754 m **PER METRE OF RUN** **PITCH** 37°

BEVELS:	COMMON RAFTER	– SEAT	37
	" "	– RIDGE	53
	HIP OR VALLEY	– SEAT	28
	" " "	– RIDGE	62
	JACK RAFTER	– EDGE	38.5
	PURLIN	– EDGE	51.5
	"	– SIDE	59

JACK RAFTERS	333 mm	**CENTRES DECREASE**	417	(in mm to 999 and thereafter in m)
	400 "	" "	501	
	500 "	" "	626	
	600 "	" "	751	

Run of Rafter	0.1	0.2	0.3	0.4	0.5	0.6	0.7	0.8	0.9	1.0
Length of Rafter	0.125	0.25	0.376	0.501	0.626	0.751	0.876	1.002	1.127	1.252
Length of Hip	0.16	0.32	0.481	0.641	0.801	0.961	1.122	1.282	1.442	1.602

RISE OF COMMON RAFTER 0.767 m **PER METRE OF RUN** **PITCH** $37\frac{1}{2}^\circ$

BEVELS:	COMMON RAFTER	– SEAT	37.5
	" "	– RIDGE	52.5
	HIP OR VALLEY	– SEAT	28.5
	" " "	– RIDGE	61.5
	JACK RAFTER	– EDGE	38.25
	PURLIN	– EDGE	51.75
	"	– SIDE	58.5

JACK RAFTERS	333 mm	**CENTRES DECREASE**	420	(in mm to 999 and thereafter in m)
	400 "	" "	505	
	500 "	" "	630	
	600 "	" "	757	

Run of Rafter	0.1	0.2	0.3	0.4	0.5	0.6	0.7	0.8	0.9	1.0
Length of Rafter	0.126	0.252	0.378	0.504	0.630	0.756	0.882	1.008	1.134	1.260
Length of Hip	0.161	0.322	0.482	0.636	0.804	1.965	1.126	1.286	1.448	1.609

RISE OF COMMON RAFTER 0.781 m **PER METRE OF RUN** **PITCH** 38°

BEVELS:	COMMON RAFTER	– SEAT	38
	" "	– RIDGE	52
	HIP OR VALLEY	– SEAT	29
	" " "	– RIDGE	61
	JACK RAFTER	– EDGE	38
	PURLIN	– EDGE	52
	"	– SIDE	58.5

JACK RAFTERS	**CENTRES DECREASE**	
333 mm	423	(in mm to 999 and thereafter in m)
400 " " "	508	
500 " " "	635	
600 " " "	761	

Run of Rafter	0.1	0.2	0.3	0.4	0.5	0.6	0.7	0.8	0.9	1.0
Length of Rafter	0.127	0.254	0.381	0.508	0.635	0.761	0.888	1.015	1.142	1.269
Length of Hip	0.162	0.323	0.485	0.646	0.808	0.969	1.131	1.293	1.454	1.616

RISE OF COMMON RAFTER 0.81 m PER METRE OF RUN   PITCH 39°

BEVELS:			
	COMMON RAFTER	– SEAT	39
	" "	– RIDGE	51
	HIP OR VALLEY	– SEAT	30
	" " "	– RIDGE	60
	JACK RAFTER	– EDGE	38
	PURLIN	– EDGE	52
	"	– SIDE	58

JACK RAFTERS			
333 mm	CENTRES DECREASE	429	(in mm to 999 and thereafter in m)
400 "	" "	515	
500 "	" "	644	
600 "	" "	772	

Run of Rafter	0.1	0.2	0.3	0.4	0.5	0.6	0.7	0.8	0.9	1.0
Length of Rafter	0.129	0.257	0.386	0.515	0.643	0.772	0.901	1.029	1.158	1.287
Length of Hip	0.163	0.326	0.489	0.652	0.815	0.978	1.141	1.304	1.467	1.63

RISE OF COMMON RAFTER 0.839 m **PER METRE OF RUN** **PITCH** 40°

BEVELS:	COMMON RAFTER	– SEAT	40
	" "	– RIDGE	50
	HIP OR VALLEY	– SEAT	30.5
	" " "	– RIDGE	59.5
	JACK RAFTER	– EDGE	37.5
	PURLIN	– EDGE	52.5
	"	– SIDE	57.5

JACK RAFTERS	333 mm	**CENTRES DECREASE**	435	(in mm to 999 and thereafter in m)
	400 "	" "	522	
	500 "	" "	653	
	600 "	" "	783	

Run of Rafter	0.1	0.2	0.3	0.4	0.5	0.6	0.7	0.8	0.9	1.0
Length of Rafter	0.131	0.261	0.392	0.522	0.653	0.783	0.914	1.044	1.175	1.305
Length of Hip	0.164	0.329	0.493	0.658	0.822	0.987	1.151	1.316	1.48	1.644

RISE OF COMMON RAFTER 0.869 m **PER METRE OF RUN** **PITCH** 41°

BEVELS:	COMMON RAFTER	– SEAT	41
	" "	– RIDGE	49
	HIP OR VALLEY	– SEAT	31.5
	" " "	– RIDGE	58.5
	JACK RAFTER	– EDGE	37
	PURLIN	– EDGE	53
	"	– SIDE	56.5

JACK RAFTERS	333 mm	**CENTRES DECREASE**	441	(in mm to 999 and thereafter in m)
	400 "	" "	530	
	500 "	" "	663	
	600 "	" "	795	

Run of Rafter	0.1	0.2	0.3	0.4	0.5	0.6	0.7	0.8	0.9	1.0
Length of Rafter	0.133	0.265	0.398	0.53	0.663	0.795	0.928	1.06	1.193	1.325
Length of Hip	0.166	0.332	0.498	0.664	0.83	0.996	1.162	1.328	1.494	1.66

RISE OF COMMON RAFTER 0.9 m **PER METRE OF RUN** **PITCH** 42°

BEVELS:	COMMON RAFTER	– SEAT	42
	" "	– RIDGE	48
	HIP OR VALLEY	– SEAT	32.5
	" " "	– RIDGE	57.5
	JACK RAFTER	– EDGE	36.5
	PURLIN	– EDGE	53.5
	"	– SIDE	56

JACK RAFTERS	333 mm	CENTRES DECREASE	448	(in mm to 999 and thereafter in m)
	400 "	" "	538	
	500 "	" "	673	
	600 "	" "	808	

Run of Rafter	0.1	0.2	0.3	0.4	0.5	0.6	0.7	0.8	0.9	1.0
Length of Rafter	0.135	0.269	0.404	0.538	0.673	0.807	0.942	1.077	1.211	1.346
Length of Hip	0.168	0.335	0.503	0.671	0.838	1.006	1.173	1.341	1.509	1.677

RISE OF COMMON RAFTER 0.916 m PER METRE OF RUN   PITCH $42\frac{1}{2}^{\circ}$

BEVELS:			
	COMMON RAFTER	– SEAT	42.5
	" "	– RIDGE	47.5
	HIP OR VALLEY	– SEAT	33
	" " "	– RIDGE	57
	JACK RAFTER	– EDGE	36.25
	PURLIN	– EDGE	53.75
	"	– SIDE	55.75

JACK RAFTERS 333 mm **CENTRES DECREASE** 452 (in mm to 999 and thereafter in m)

JACK RAFTERS	333 mm	**CENTRES DECREASE**	452
	400 "	" "	543
	500 "	" "	679
	600 "	" "	815

Run of Rafter	0.1	0.2	0.3	0.4	0.5	0.6	0.7	0.8	0.9	1.0
Length of Rafter	0.137	0.271	0.406	0.543	0.678	0.814	0.949	1.085	1.221	1.356
Length of Hip	0.170	0.337	0.505	0.674	0.842	1.011	1.179	1.348	1.569	1.685

RISE OF COMMON RAFTER 0.933 m **PER METRE OF RUN** **PITCH** 43°

BEVELS:	COMMON RAFTER	– SEAT	43
	" "	– RIDGE	47
	HIP OR VALLEY	– SEAT	33.5
	" " "	– RIDGE	56.5
	JACK RAFTER	– EDGE	36
	PURLIN	– EDGE	54
	"	– SIDE	55.5

JACK RAFTERS	333 mm	**CENTRES**	**DECREASE**	455	(in mm to 999 and
	400 "	"	"	547	thereafter in m)
	500 "	"	"	684	
	600 "	"	"	820	

Run of Rafter	0.1	0.2	0.3	0.4	0.5	0.6	0.7	0.8	0.9	1.0
Length of Rafter	0.137	0.273	0.41	0.547	0.684	0.82	0.967	1.094	1.231	1.367
Length of Hip	0.169	0.339	0.508	0.678	0.847	1.016	1.186	1.355	1.525	1.694

RISE OF COMMON RAFTER 0.966 m **PER METRE OF RUN** **PITCH** 44°

BEVELS:	COMMON RAFTER	– SEAT	44
	" "	– RIDGE	46
	HIP OR VALLEY	– SEAT	34.5
	" " "	– RIDGE	55.5
	JACK RAFTER	– EDGE	35.5
	PURLIN	– EDGE	54.5
	"	– SIDE	55

JACK RAFTERS	333 mm	**CENTRES**	**DECREASE**	463	(in mm to 999 and thereafter in m)
	400 "	"	"	556	
	500 "	"	"	695	
	600 "	"	"	834	

Run of Rafter	0.1	0.2	0.3	0.4	0.5	0.6	0.7	0.8	0.9	1.0
Length of Rafter	0.139	0.278	0.417	0.556	0.695	0.834	0.973	1.111	1.251	1.39
Length of Hip	0.171	0.342	0.514	0.685	0.856	1.027	1.199	1.37	1.541	1.712

RISE OF COMMON RAFTER 1.0 m **PER METRE OF RUN** **PITCH** 45°

BEVELS:			
	COMMON RAFTER	– SEAT	45
	″ ″	– RIDGE	45
	HIP OR VALLEY	– SEAT	35.5
	″ ″ ″	– RIDGE	54.5
	JACK RAFTER	– EDGE	35.5
	PURLIN	– EDGE	54.5
	″	– SIDE	54.5

JACK RAFTERS			
333 mm	**CENTRES DECREASE**	471	(in mm to 999 and thereafter in m)
400 ″	″ ″	566	
500 ″	″ ″	707	
600 ″	″ ″	848	

Run of Rafter	0.1	0.2	0.3	0.4	0.5	0.6	0.7	0.8	0.9	1.0
Length of Rafter	0.141	0.283	0.424	0.566	0.707	0.848	0.99	1.131	1.273	1.414
Length of Hip	0.173	0.346	0.519	0.693	0.866	1.039	1.212	1.386	1.559	1.732

RISE OF COMMON RAFTER 1.036 m **PER METRE OF RUN** **PITCH** 46°

BEVELS:			
	COMMON RAFTER	– SEAT	46
	" "	– RIDGE	44
	HIP OR VALLEY	– SEAT	36
	" " "	– RIDGE	54
	JACK RAFTER	– EDGE	35
	PURLIN	– EDGE	55
	"	– SIDE	54.5

JACK RAFTERS				
333 mm	**CENTRES**	**DECREASE**	480	(in mm to 999 and
400 "	"	"	576	thereafter in m)
500 "	"	"	720	
600 "	"	"	864	

Run of Rafter	0.1	0.2	0.3	0.4	0.5	0.6	0.7	0.8	0.9	1.0
Length of Rafter	0.144	0.288	0.432	0.576	0.72	0.864	1.058	1.152	1.296	1.44
Length of Hip	0.175	0.351	0.526	0.701	0.876	1.052	1.227	1.402	1.578	1.753

RISE OF COMMON RAFTER 1.072 m PER METRE OF RUN   PITCH 47°

BEVELS:			
	COMMON RAFTER	– SEAT	47
	" "	– RIDGE	43
	HIP OR VALLEY	– SEAT	37
	" " "	– RIDGE	53
	JACK RAFTER	– EDGE	34.5
	PURLIN	– EDGE	55.5
	"	– SIDE	54

JACK RAFTERS			
333 mm	CENTRES DECREASE	488	(in mm to 999 and thereafter in m)
400 "	" "	586	
500 "	" "	733	
600 "	" "	880	

Run of Rafter	0.1	0.2	0.3	0.4	0.5	0.6	0.7	0.8	0.9	1.0
Length of Rafter	0.147	0.293	0.44	0.587	0.733	0.88	1.026	1.173	1.32	1.466
Length of Hip	0.177	0.355	0.532	0.71	0.887	1.065	1.242	1.42	1.597	1.775

RISE OF COMMON RAFTER 1.111 m PER METRE OF RUN — PITCH 48°

BEVELS:			
	COMMON RAFTER	– SEAT	48
	" "	– RIDGE	42
	HIP OR VALLEY	– SEAT	38
	" " "	– RIDGE	52
	JACK RAFTER	– EDGE	34
	PURLIN	– EDGE	56
	"	– SIDE	53.5

JACK RAFTERS					
JACK RAFTERS	333 mm	**CENTRES DECREASE**	498	(in mm to 999 and thereafter in m)
	400 "	" "	598	
	500 "	" "	747	
	600 "	" "	896	

Run of Rafter	0.1	0.2	0.3	0.4	0.5	0.6	0.7	0.8	0.9	1.0
Length of Rafter	0.149	0.299	0.448	0.598	0.747	0.897	1.046	1.196	1.345	1.494
Length of Hip	0.18	0.36	0.539	0.719	0.899	1.079	1.259	1.438	1.618	1.798

RISE OF COMMON RAFTER 1.15 m PER METRE OF RUN   PITCH 49°

BEVELS:	COMMON RAFTER	– SEAT	49
	" "	– RIDGE	41
	HIP OR VALLEY	– SEAT	39
	" " "	– RIDGE	51
	JACK RAFTER	– EDGE	33.5
	PURLIN	– EDGE	56.5
	"	– SIDE	53

JACK RAFTERS	333 mm	CENTRES DECREASE	508	(in mm to 999 and thereafter in m)
	400 "	" "	610	
	500 "	" "	762	
	600 "	" "	914	

Run of Rafter	0.1	0.2	0.3	0.4	0.5	0.6	0.7	0.8	0.9	1.0
Length of Rafter	0.152	0.305	0.457	0.61	0.762	0.915	1.067	1.219	1.372	1.524
Length of Hip	0.182	0.365	0.547	0.729	0.912	1.094	1.276	1.458	1.641	1.823

RISE OF COMMON RAFTER 1.192 m PER METRE OF RUN   PITCH 50°

BEVELS:			
	COMMON RAFTER	– SEAT	50
	" "	– RIDGE	40
	HIP OR VALLEY	– SEAT	40
	" " "	– RIDGE	50
	JACK RAFTER	– EDGE	32.5
	PURLIN	– EDGE	57.5
	"	– SIDE	52.5

JACK RAFTERS			
333 mm	**CENTRES DECREASE**	518	(in mm to 999 and thereafter in m)
400 "	" "	622	
500 "	" "	778	
600 "	" "	934	

Run of Rafter	0.1	0.2	0.3	0.4	0.5	0.6	0.7	0.8	0.9	1.0
Length of Rafter	0.156	0.311	0.467	0.622	0.778	0.933	1.089	1.246	1.4	1.556
Length of Hip	0.185	0.37	0.555	0.74	0.925	1.11	1.295	1.48	1.664	1.849

RISE OF COMMON RAFTER 1.235 m **PER METRE OF RUN** **PITCH** 51°

BEVELS:	COMMON RAFTER	– SEAT	51
	" "	– RIDGE	39
	HIP OR VALLEY	– SEAT	41
	" " "	– RIDGE	49
	JACK RAFTER	– EDGE	32
	PURLIN	– EDGE	58
	"	– SIDE	52

JACK RAFTERS	333 mm	**CENTRES DECREASE**	529	(in mm to 999 and thereafter in m)
	400 "	" "	636	
	500 "	" "	795	
	600 "	" "	953	

Run of Rafter	0.1	0.2	0.3	0.4	0.5	0.6	0.7	0.8	0.9	1.0
Length of Rafter	0.159	0.318	0.477	0.636	0.795	0.953	1.012	1.271	1.43	1.589
Length of Hip	0.188	0.375	0.563	0.751	0.936	1.126	1.314	1.502	1.69	1.877

RISE OF COMMON RAFTER 1.28 m PER METRE OF RUN — PITCH 52°

BEVELS:			
COMMON RAFTER	– SEAT	52
〃 〃	– RIDGE	38
HIP OR VALLEY	– SEAT	42
〃 〃 〃	– RIDGE	48
JACK RAFTER	– EDGE	31.5
PURLIN	– EDGE	58.5
〃	– SIDE	52

JACK RAFTERS	CENTRES DECREASE	
333 mm	541	(in mm to 999 and thereafter in m)
400 〃	650	
500 〃	812	
600 〃	974	

Run of Rafter	0.1	0.2	0.3	0.4	0.5	0.6	0.7	0.8	0.9	1.0
Length of Rafter	0.162	0.325	0.487	0.65	0.812	0.974	1.137	1.299	1.462	1.624
Length of Hip	0.191	0.381	0.572	0.763	0.954	1.144	1.335	1.526	1.717	1.907

RISE OF COMMON RAFTER 1.327 m PER METRE OF RUN   PITCH 53°

BEVELS:			
	COMMON RAFTER	– SEAT	53
	" "	– RIDGE	37
	HIP OR VALLEY	– SEAT	43
	" " "	– RIDGE	47
	JACK RAFTER	– EDGE	31
	PURLIN	– EDGE	59
	"	– SIDE	51.5

JACK RAFTERS			
333 mm	CENTRES DECREASE	553	(in mm to 999 and thereafter in m)
400 "	" "	665	
500 "	" "	831	
600 "	" "	997	

Run of Rafter	0.1	0.2	0.3	0.4	0.5	0.6	0.7	0.8	0.9	1.0
Length of Rafter	0.166	0.332	0.498	0.665	0.831	0.997	1.163	1.329	1.495	1.662
Length of Hip	0.194	0.388	0.582	0.776	0.97	1.164	1.358	1.551	1.745	1.939

RISE OF COMMON RAFTER 1.376 m **PER METRE OF RUN** **PITCH** 54°

BEVELS:	COMMON RAFTER	– SEAT	54
	" "	– RIDGE	36
	HIP OR VALLEY	– SEAT	44
	" " "	– RIDGE	46
	JACK RAFTER	– EDGE	30.5
	PURLIN	– EDGE	59.5
	"	– SIDE	51

JACK RAFTERS	333 mm	**CENTRES**	**DECREASE**	567	(in mm to 999 and
	400 "	"	"	680	thereafter in m)
	500 "	"	"	850	
	600 "	"	"	1.021	

Run of Rafter	0.1	0.2	0.3	0.4	0.5	0.6	0.7	0.8	0.9	1.0
Length of Rafter	0.17	0.34	0.51	0.681	0.851	1.021	1.191	1.361	1.531	1.701
Length of Hip	0.197	0.395	0.592	0.789	0.987	1.184	1.381	1.579	1.776	1.973

RISE OF COMMON RAFTER 1.428 m PER METRE OF RUN — PITCH 55°

BEVELS:			
	COMMON RAFTER	– SEAT	55
	" "	– RIDGE	35
	HIP OR VALLEY	– SEAT	45.5
	" " "	– RIDGE	44.5
	JACK RAFTER	– EDGE	30
	PURLIN	– EDGE	60
	"	– SIDE	50.5

JACK RAFTERS			
333 mm	CENTRES DECREASE	580	(in mm to 999 and thereafter in m)
400 "	" "	697	
500 "	" "	872	
600 "	" "	1.046	

Run of Rafter	0.1	0.2	0.3	0.4	0.5	0.6	0.7	0.8	0.9	1.0
Length of Rafter	0.174	0.349	0.523	0.697	0.872	1.046	1.22	1.395	1.569	1.743
Length of Hip	0.201	0.402	0.603	0.804	1.005	1.206	1.407	1.608	1.809	2.01

RISE OF COMMON RAFTER 1.5 m **PER METRE OF RUN** **PITCH** 56° 18′
(Italian pitch)

BEVELS:	COMMON RAFTER	– SEAT	56.5
	" "	– RIDGE	33.5
	HIP OR VALLEY	– SEAT	46.5
	" " "	– RIDGE	43.5
	JACK RAFTER	– EDGE	29
	PURLIN	– EDGE	61
	"	– SIDE	50

JACK RAFTERS	333 mm	**CENTRES DECREASE**	600	(in mm to 999 and thereafter in m)
	400 "	" "	720	
	500 "	" "	900	
	600 "	" "	1.082	

Run of Rafter	0.1	0.2	0.3	0.4	0.5	0.6	0.7	0.8	0.9	1.0
Length of Rafter	0.18	0.361	0.541	0.721	0.902	1.082	1.262	1.442	1.623	1.803
Length of Hip	0.206	0.412	0.618	0.824	1.031	1.237	1.443	1.649	1.855	2.061

RISE OF COMMON RAFTER 1.6 m PER METRE OF RUN — PITCH 58°

BEVELS:			
	COMMON RAFTER	– SEAT	58
	" "	– RIDGE	32
	HIP OR VALLEY	– SEAT	48.5
	" " "	– RIDGE	41.5
	JACK RAFTER	– EDGE	28
	PURLIN	– EDGE	62
	"	– SIDE	49.5

JACK RAFTERS			
333 mm	**CENTRES DECREASE**	628	(in mm to 999 and thereafter in m)
400 "	" "	755	
500 "	" "	944	
600 "	" "	1.132	

Run of Rafter	0.1	0.2	0.3	0.4	0.5	0.6	0.7	0.8	0.9	1.0
Length of Rafter	0.189	0.377	0.566	0.755	0.944	1.132	1.321	1.51	1.698	1.887
Length of Hip	0.214	0.428	0.641	0.855	1.069	1.283	1.496	1.71	1.924	2.136

RISE OF COMMON RAFTER 1.664 m PER METRE OF RUN   PITCH 59°

BEVELS:			
	COMMON RAFTER	– SEAT	59
	" "	– RIDGE	31
	HIP OR VALLEY	– SEAT	49.5
	" " "	– RIDGE	40.5
	JACK RAFTER	– EDGE	27.5
	PURLIN	– EDGE	62.5
	"	– SIDE	49.5

JACK RAFTERS			
333 mm	**CENTRES DECREASE**	647	(in mm to 999 and thereafter in m)
400 "	" "	777	
500 "	" "	971	
600 "	" "	1.165	

Run of Rafter	0.1	0.2	0.3	0.4	0.5	0.6	0.7	0.8	0.9	1.0
Length of Rafter	0.194	0.398	0.582	0.777	0.971	1.165	1.359	1.553	1.747	1.942
Length of Hip	0.218	0.437	0.655	0.874	1.092	1.31	1.529	1.747	1.965	2.184

RISE OF COMMON RAFTER 1.732 m **PER METRE OF RUN** **PITCH** 60°
(Equilateral pitch)

BEVELS:			
	COMMON RAFTER	– SEAT	60
	" "	– RIDGE	30
	HIP OR VALLEY	– SEAT	51
	" " "	– RIDGE	39
	JACK RAFTER	– EDGE	26.5
	PURLIN	– EDGE	63.5
	"	– SIDE	49

JACK RAFTERS				
	333 mm	**CENTRES DECREASE**	666	(in mm to 999 and thereafter in m)
	400 "	" "	800	
	500 "	" "	1.000	
	600 "	" "	1.200	

Run of Rafter	0.1	0.2	0.3	0.4	0.5	0.6	0.7	0.8	0.9	1.0
Length of Rafter	0.2	0.4	0.6	0.8	1.0	1.2	1.4	1.6	1.8	2.0
Length of Hip	0.224	0.447	0.671	0.894	1.118	1.342	1.565	1.789	2.012	2.236

RISE OF COMMON RAFTER 1.804 m **PER METRE OF RUN** **PITCH** 61°

BEVELS:	COMMON RAFTER	– SEAT	61
	" "	– RIDGE	29
	HIP OR VALLEY	– SEAT	52
	" " "	– RIDGE	38
	JACK RAFTER	– EDGE	26
	PURLIN	– EDGE	64
	"	– SIDE	49

JACK RAFTERS				
333 mm	**CENTRES**	**DECREASE**	687	(in mm to 999 and
400 "	"	"	825	thereafter in m)
500 "	"	"	1.032	
600 "	"	"	1.238	

Run of Rafter	0.1	0.2	0.3	0.4	0.5	0.6	0.7	0.8	0.9	1.0
Length of Rafter	0.206	0.413	0.619	0.825	1.031	1.238	1.444	1.65	1.857	2.063
Length of Hip	0.229	0.458	0.688	0.917	1.146	1.375	1.605	1.834	2.063	2.292

RISE OF COMMON RAFTER 1.88 m **PER METRE OF RUN** **PITCH** 62°

BEVELS:			
	COMMON RAFTER	– SEAT	62
	" "	– RIDGE	28
	HIP OR VALLEY	– SEAT	53
	" " "	– RIDGE	37
	JACK RAFTER	– EDGE	25
	PURLIN	– EDGE	65
	"	– SIDE	48.5

JACK RAFTERS			
333 mm	**CENTRES DECREASE**	709	(in mm to 999 and thereafter in m)
400 "	" "	852	
500 "	" "	1.065	
600 "	" "	1.278	

Run of Rafter	0.1	0.2	0.3	0.4	0.5	0.6	0.7	0.8	0.9	1.0
Length of Rafter	0.213	0.426	0.639	0.852	1.065	1.278	1.491	1.704	1.917	2.13
Length of Hip	0.235	0.471	0.706	0.941	1.177	1.412	1.647	1.882	2.118	2.353

RISE OF COMMON RAFTER 2.0 m **PER METRE OF RUN** **PITCH** 63° 26′ (Gothic pitch)

BEVELS:	COMMON RAFTER	– SEAT	63.5
	" "	– RIDGE	26.5
	HIP OR VALLEY	– SEAT	54.5
	" " "	– RIDGE	35.5
	JACK RAFTER	– EDGE	24
	PURLIN	– EDGE	66
	"	– SIDE	48

JACK RAFTERS	333 mm **CENTRES DECREASE**	745	(in mm to 999 and thereafter in m)
	400 " " "	894	
	500 " " "	1.118	
	600 " " "	1.342	

Run of Rafter	0.1	0.2	0.3	0.4	0.5	0.6	0.7	0.8	0.9	1.0
Length of Rafter	0.224	0.447	0.671	0.894	1.118	1.342	1.565	1.789	2.012	2.236
Length of Hip	0.245	0.49	0.735	0.98	1.225	1.47	1.715	1.96	2.205	2.45

RISE OF COMMON RAFTER 2.145 m **PER METRE OF RUN** **PITCH** 65°

BEVELS:	COMMON RAFTER	– SEAT	65
	" "	– RIDGE	25
	HIP OR VALLEY	– SEAT	56.5
	" " "	– RIDGE	33.5
	JACK RAFTER	– EDGE	23
	PURLIN	– EDGE	67
	"	– SIDE	48

JACK RAFTERS	333 mm	**CENTRES DECREASE**	788	(in mm to 999 and thereafter in m)
	400 "	" "	946	
	500 "	" "	1.183	
	600 "	" "	1.420	

Run of Rafter	0.1	0.2	0.3	0.4	0.5	0.6	0.7	0.8	0.9	1.0
Length of Rafter	0.237	0.473	0.71	0.946	1.183	1.42	1.656	1.893	2.13	2.366
Length of Hip	0.257	0.514	0.771	1.028	1.284	1.541	1.798	2.055	2.312	2.569

RISE OF COMMON RAFTER 2.246 m **PER METRE OF RUN** **PITCH** 66°

BEVELS:	COMMON RAFTER	– SEAT	66
	" "	– RIDGE	24
	HIP OR VALLEY	– SEAT	58
	" " "	– RIDGE	32
	JACK RAFTER	– EDGE	22
	PURLIN	– EDGE	68
	"	– SIDE	47.5

JACK RAFTERS	333 mm	**CENTRES**	**DECREASE**	819	(in mm to 999 and thereafter in m)
	400 "	"	"	984	
	500 "	"	"	1.230	
	600 "	"	"	1.475	

Run of Rafter	0.1	0.2	0.3	0.4	0.5	0.6	0.7	0.8	0.9	1.0
Length of Rafter	0.246	0.492	0.738	0.983	1.229	1.475	1.721	1.967	2.213	2.459
Length of Hip	0.265	0.531	0.796	1.062	1.327	1.593	1.858	2.123	2.389	2.654

RISE OF COMMON RAFTER 2.356 m **PER METRE OF RUN** **PITCH** 67°

BEVELS:	COMMON RAFTER	– SEAT	67
	" "	– RIDGE	23
	HIP OR VALLEY	– SEAT	59
	" " "	– RIDGE	31
	JACK RAFTER	– EDGE	21.5
	PURLIN	– EDGE	68.5
	"	– SIDE	47.5

JACK RAFTERS	333 mm **CENTRES DECREASE**	852	(in mm to 999 and thereafter in m)
	400 " " "	1.024	
	500 " " "	1.280	
	600 " " "	1.535	

Run of Rafter	0.1	0.2	0.3	0.4	0.5	0.6	0.7	0.8	0.9	1.0
Length of Rafter	0.256	0.512	0.768	1.024	1.28	1.536	1.792	2.047	2.303	2.559
Length of Hip	0.275	0.55	0.824	1.099	1.374	1.649	1.922	2.198	2.473	2.748

RISE OF COMMON RAFTER 2.475 m PER METRE OF RUN — PITCH 68°

BEVELS:			
	COMMON RAFTER	– SEAT	68
	" "	– RIDGE	22
	HIP OR VALLEY	– SEAT	60.5
	" " "	– RIDGE	29.5
	JACK RAFTER	– EDGE	20.5
	PURLIN	– EDGE	69.5
	"	– SIDE	47

JACK RAFTERS			
333 mm	CENTRES DECREASE	889	(in mm to 999 and thereafter in m)
400 "	" "	1.068	
500 "	" "	1.335	
600 "	" "	1.601	

Run of Rafter	0.1	0.2	0.3	0.4	0.5	0.6	0.7	0.8	0.9	1.0
Length of Rafter	0.267	0.534	0.801	1.068	1.335	1.602	1.869	2.136	2.403	2.669
Length of Hip	0.285	0.57	0.855	1.14	1.425	1.71	1.995	2.28	2.566	2.851

RISE OF COMMON RAFTER 2.605 m PER METRE OF RUN — PITCH 69°

BEVELS:			
COMMON RAFTER	–	SEAT	69
〃 〃	–	RIDGE	21
HIP OR VALLEY	–	SEAT	61.5
〃 〃 〃	–	RIDGE	28.5
JACK RAFTER	–	EDGE	19.5
PURLIN	–	EDGE	70.5
〃	–	SIDE	47

JACK RAFTERS	CENTRES DECREASE	
333 mm	930	(in mm to 999 and thereafter in m)
400 〃	1.116	
500 〃	1.395	
600 〃	1.674	

Run of Rafter	0.1	0.2	0.3	0.4	0.5	0.6	0.7	0.8	0.9	1.0
Length of Rafter	0.279	0.558	0.837	1.116	1.395	1.674	1.953	2.232	2.511	2.79
Length of Hip	0.296	0.593	0.889	1.186	1.482	1.779	2.075	2.371	2.688	2.964

RISE OF COMMON RAFTER 2.747 m PER METRE OF RUN PITCH 70°

BEVELS:			
	COMMON RAFTER	– SEAT	70
	" "	– RIDGE	20
	HIP OR VALLEY	– SEAT	63
	" " "	– RIDGE	27
	JACK RAFTER	– EDGE	19
	PURLIN	– EDGE	71
	"	– SIDE	47

JACK RAFTERS			
333 mm	CENTRES DECREASE	975	(in mm to 999 and thereafter in m)
400 "	" "	1.170	
500 "	" "	1.462	
600 "	" "	1.754	

Run of Rafter	0.1	0.2	0.3	0.4	0.5	0.6	0.7	0.8	0.9	1.0
Length of Rafter	0.292	0.585	0.877	1.17	1.462	1.754	2.047	2.339	2.631	2.924
Length of Hip	0.309	0.618	0.927	1.236	1.545	1.854	2.163	2.472	2.781	3.09

RISE OF COMMON RAFTER 2.904 m PER METRE OF RUN — PITCH 71°

BEVELS:			
	COMMON RAFTER	– SEAT	71
	" "	– RIDGE	19
	HIP OR VALLEY	– SEAT	64
	" " "	– RIDGE	26
	JACK RAFTER	– EDGE	18
	PURLIN	– EDGE	72
	"	– SIDE	46.5

JACK RAFTERS			
333 mm	**CENTRES DECREASE**	1.024	(in mm to 999 and
400 "	" "	1.229	thereafter in m)
500 "	" "	1.536	
600 "	" "	1.843	

Run of Rafter	0.1	0.2	0.3	0.4	0.5	0.6	0.7	0.8	0.9	1.0
Length of Rafter	0.307	0.614	0.921	1.229	1.536	1.843	2.15	2.457	2.764	3.072
Length of Hip	0.323	0.646	0.969	1.292	1.615	1.938	2.261	2.584	2.907	3.23

RISE OF COMMON RAFTER 3.078 m PER METRE OF RUN   PITCH 72°

BEVELS:	COMMON RAFTER	– SEAT	72
	" "	– RIDGE	18
	HIP OR VALLEY	– SEAT	65.5
	" " "	– RIDGE	24.5
	JACK RAFTER	– EDGE	17
	PURLIN	– EDGE	73
	"	– SIDE	46.5

JACK RAFTERS	333 mm	CENTRES DECREASE	1.078	(in mm to 999 and
	400 "	" "	1.294	thereafter in m)
	500 "	" "	1.618	
	600 "	" "	1.942	

Run of Rafter	0.1	0.2	0.3	0.4	0.5	0.6	0.7	0.8	0.9	1.0
Length of Rafter	0.324	0.647	0.971	1.294	1.618	1.942	2.266	2.589	2.912	3.236
Length of Hip	0.339	0.677	1.016	1.355	1.694	2.032	2.371	2.71	3.048	3.387

RISE OF COMMON RAFTER 3.271 m PER METRE OF RUN — PITCH 73°

BEVELS:			
	COMMON RAFTER	– SEAT	73
	" "	– RIDGE	17
	HIP OR VALLEY	– SEAT	66.5
	" " "	– RIDGE	23.5
	JACK RAFTER	– EDGE	16.5
	PURLIN	– EDGE	73.5
	"	– SIDE	46.5

JACK RAFTERS			
333 mm	**CENTRES DECREASE**	1.140	(in mm to 999 and thereafter in m)
400 "	" "	1.368	
500 "	" "	1.710	
600 "	" "	2.052	

Run of Rafter	0.1	0.2	0.3	0.4	0.5	0.6	0.7	0.8	0.9	1.0
Length of Rafter	0.342	0.684	1.026	1.368	1.71	2.052	2.394	2.736	3.078	3.42
Length of Hip	0.356	0.713	1.069	1.425	1.782	2.138	2.494	2.851	3.207	3.563

RISE OF COMMON RAFTER 3.487 m **PER METRE OF RUN** **PITCH** 74°

BEVELS:	COMMON RAFTER	– SEAT	74
	" "	– RIDGE	16
	HIP OR VALLEY	– SEAT	68
	" " "	– RIDGE	22
	JACK RAFTER	– EDGE	15.5
	PURLIN	– EDGE	74.5
	"	– SIDE	46

JACK RAFTERS	333 mm	**CENTRES**	**DECREASE**	1.209	(in mm to 999 and thereafter in m)
	400 "	"	"	1.451	
	500 "	"	"	1.814	
	600 "	"	"	2.177	

Run of Rafter	0.1	0.2	0.3	0.4	0.5	0.6	0.7	0.8	0.9	1.0
Length of Rafter	0.363	0.726	1.088	1.451	1.814	2.177	2.54	2.902	3.265	3.628
Length of Hip	0.376	0.753	1.129	1.505	1.882	2.258	2.634	3.011	3.387	3.763

RISE OF COMMON RAFTER 3.732 m **PER METRE OF RUN** **PITCH** 75°

BEVELS:	COMMON RAFTER	– SEAT	75
	" "	– RIDGE	15
	HIP OR VALLEY	– SEAT	69
	" " "	– RIDGE	21
	JACK RAFTER	– EDGE	14.5
	PURLIN	– EDGE	75.5
	"	– SIDE	46

JACK RAFTERS	333 mm	**CENTRES**	**DECREASE**	1.289	(in mm to 999 and
	400 "	"	"	1.546	thereafter in m)
	500 "	"	"	1.932	
	600 "	"	"	2.318	

Run of Rafter	0.1	0.2	0.3	0.4	0.5	0.6	0.7	0.8	0.9	1.0
Length of Rafter	0.386	0.773	1.159	1.545	1.932	2.318	2.705	3.091	3.477	3.864
Length of Hip	0.399	0.789	1.197	1.596	1.996	2.395	2.794	3.193	3.592	3.991

5 IMPERIAL CALCULATION TABLES

5° PITCH

RISE OF COMMON RAFTER $1\frac{1}{16}''$ PER FOOT OF RUN

BEVELS:

1.	COMMON RAFTER	–	SEAT	5
2.	" "	–	RIDGE	85
3.	HIP OR VALLEY	–	SEAT	
4.	" " "	–	RIDGE	
5.	JACK RAFTER	–	EDGE	
6.	PURLIN	–	EDGE	
7.	"	–	SIDE	

JACK RAFTERS 16 in. CENTRES DECREASE

RUN OF RAFTER *ins.*	$\frac{1}{2}$	1	2	3	4	5	6	7	8	9	10	11
LENGTH OF RAFTER												
LENGTH OF HIP												

RUN OF RAFTER *ft.*	1	2	3	4	5	6	7	8	9	10
LENGTH OF RAFTER	1·0	2·0	$3{\cdot}0\frac{1}{8}$	$4{\cdot}0\frac{1}{8}$	$5{\cdot}0\frac{1}{4}$	$6{\cdot}0\frac{1}{4}$	$7{\cdot}0\frac{1}{4}$	$8{\cdot}0\frac{3}{8}$	$9{\cdot}0\frac{3}{8}$	$10{\cdot}0\frac{1}{2}$
LENGTH OF HIP										

6° PITCH

RISE OF COMMON RAFTER 1¼″ PER FOOT OF RUN

BEVELS:

1.	COMMON RAFTER	– SEAT	6
2.	″ ″	– RIDGE	84
3.	HIP OR VALLEY	– SEAT	
4.	″ ″ ″	– RIDGE	
5.	JACK RAFTER	– EDGE	
6.	PURLIN	– EDGE	
7.	″	– SIDE	

JACK RAFTERS 16 in. CENTRES DECREASE

RUN OF RAFTER *ins.*	½	1	2	3	4	5	6	7	8	9	10	11
LENGTH OF RAFTER												
LENGTH OF HIP												

RUN OF RAFTER *ft.*	1	2	3	4	5	6	7	8	9	10
LENGTH OF RAFTER	1·0	2·0⅛	3·0⅛	4·0¼	5·0¼	6·0⅜	7·0½	8·0½	9·0⅝	10·0⅝
LENGTH OF HIP										

7° PITCH

RISE OF COMMON RAFTER $1\frac{1}{2}''$ PER FOOT OF RUN

BEVELS:				
1.	COMMON RAFTER	–	SEAT	7
2.	" "	–	RIDGE	83
3.	HIP OR VALLEY	–	SEAT	
4.	" " "	–	RIDGE	
5.	JACK RAFTER	–	EDGE	
6.	PURLIN	–	EDGE	
7.	"	–	SIDE	

JACK RAFTERS 16 in. CENTRES DECREASE

RUN OF RAFTER *ins.*	$\frac{1}{2}$	1	2	3	4	5	6	7	8	9	10	11
LENGTH OF RAFTER												
LENGTH OF HIP												

RUN OF RAFTER *ft.*	1	2	3	4	5	6	7	8	9	10
LENGTH OF RAFTER	1·0	2·0$\frac{1}{8}$	3·0$\frac{1}{4}$	4·0$\frac{1}{4}$	5·0$\frac{3}{8}$	6·0$\frac{1}{2}$	7·0$\frac{5}{8}$	8·0$\frac{5}{8}$	9·0$\frac{3}{4}$	10·0$\frac{7}{8}$
LENGTH OF HIP										

8° PITCH

RISE OF COMMON RAFTER $1\frac{11}{16}''$ PER FOOT OF RUN

BEVELS:				
	1.	COMMON RAFTER	– SEAT	8
	2.	" "	– RIDGE	82
	3.	HIP OR VALLEY	– SEAT	
	4.	" " "	– RIDGE	
	5.	JACK RAFTER	– EDGE	
	6.	PURLIN	– EDGE	
	7.	"	– SIDE	

JACK RAFTERS 16 in. CENTRES DECREASE

RUN OF RAFTER *ins.*	$\frac{1}{2}$	1	2	3	4	5	6	7	8	9	10	11
LENGTH OF RAFTER												
LENGTH OF HIP												

RUN OF RAFTER *ft.*	1	2	3	4	5	6	7	8	9	10
LENGTH OF RAFTER	$1{\cdot}0\frac{1}{8}$	$2{\cdot}0\frac{1}{4}$	$3{\cdot}0\frac{3}{8}$	$4{\cdot}0\frac{1}{2}$	$5{\cdot}0\frac{5}{8}$	$6{\cdot}0\frac{3}{4}$	$7{\cdot}0\frac{7}{8}$	8·1	9·1	$10{\cdot}1\frac{1}{8}$
LENGTH OF HIP										

9° PITCH

RISE OF COMMON RAFTER $1\frac{7}{8}''$ PER FOOT OF RUN

BEVELS:				
1.	COMMON RAFTER	–	SEAT	9
2.	″ ″	–	RIDGE	81
3.	HIP OR VALLEY	–	SEAT	
4.	″ ″ ″	–	RIDGE	
5.	JACK RAFTER	–	EDGE	
6.	PURLIN	–	EDGE	
7.	″	–	SIDE	

JACK RAFTERS 16 in. CENTRES DECREASE

RUN OF RAFTER *ins.*	$\frac{1}{2}$	1	2	3	4	5	6	7	8	9	10	11
LENGTH OF RAFTER												
LENGTH OF HIP												

RUN OF RAFTER *ft.*	1	2	3	4	5	6	7	8	9	10
LENGTH OF RAFTER	$1{\cdot}0\frac{1}{8}$	$2{\cdot}0\frac{1}{4}$	$3{\cdot}0\frac{3}{8}$	$4{\cdot}0\frac{5}{8}$	$5{\cdot}0\frac{3}{4}$	$6{\cdot}0\frac{7}{8}$	$7{\cdot}1$	$8{\cdot}1\frac{1}{4}$	$9{\cdot}1\frac{3}{8}$	$10{\cdot}1\frac{1}{2}$
LENGTH OF HIP										

10° PITCH

RISE OF COMMON RAFTER $2\frac{1}{8}''$ PER FOOT OF RUN

BEVELS: 1. COMMON RAFTER – SEAT 10
2. " " – RIDGE 80
3. HIP OR VALLEY – SEAT
4. " " " – RIDGE
5. JACK RAFTER – EDGE
6. PURLIN – EDGE
7. " – SIDE

JACK RAFTERS 16 in. CENTRES DECREASE

RUN OF RAFTER *ins.*	$\frac{1}{2}$	1	2	3	4	5	6	7	8	9	10	11
LENGTH OF RAFTER												
LENGTH OF HIP												

RUN OF RAFTER *ft.*	1	2	3	4	5	6	7	8	9	10
LENGTH OF RAFTER	$1{\cdot}0\frac{1}{8}$	$2{\cdot}0\frac{3}{8}$	$3{\cdot}0\frac{1}{2}$	$4{\cdot}0\frac{3}{4}$	5·1	$6{\cdot}1\frac{1}{8}$	$7{\cdot}1\frac{1}{4}$	$8{\cdot}1\frac{1}{2}$	$9{\cdot}1\frac{5}{8}$	$10{\cdot}1\frac{7}{8}$
LENGTH OF HIP										

11° PITCH

RISE OF COMMON RAFTER $2\frac{5}{16}$″ PER FOOT OF RUN

BEVELS:

1.	COMMON RAFTER	– SEAT	11
2.	″ ″	– RIDGE	79
3.	HIP OR VALLEY	– SEAT	
4.	″ ″ ″	– RIDGE	
5.	JACK RAFTER	– EDGE	
6.	PURLIN	– EDGE	
7.	″	– SIDE	

JACK RAFTERS 16 in. CENTRES DECREASE

RUN OF RAFTER *ins.*	$\frac{1}{2}$	1	2	3	4	5	6	7	8	9	10	11
LENGTH OF RAFTER												
LENGTH OF HIP												

RUN OF RAFTER *ft.*	1	2	3	4	5	6	7	8	9	10
LENGTH OF RAFTER	$1{\cdot}0\frac{1}{4}$	$2{\cdot}0\frac{1}{2}$	$3{\cdot}0\frac{3}{4}$	4·1	$5{\cdot}1\frac{1}{8}$	$6{\cdot}1\frac{1}{4}$	$7{\cdot}1\frac{1}{2}$	$8{\cdot}1\frac{3}{4}$	9·2	$10{\cdot}2\frac{1}{4}$
LENGTH OF HIP										

12° PITCH

RISE OF COMMON RAFTER $2\frac{9}{16}''$ PER FOOT OF RUN

BEVELS:				
	1. COMMON RAFTER	–	SEAT	12
	2. 〃 〃	–	RIDGE	78
	3. HIP OR VALLEY	–	SEAT	
	4. 〃 〃 〃	–	RIDGE	
	5. JACK RAFTER	–	EDGE	
	6. PURLIN	–	EDGE	
	7. 〃	–	SIDE	

JACK RAFTERS 16 in. CENTRES DECREASE

RUN OF RAFTER ins.	$\frac{1}{2}$	1	2	3	4	5	6	7	8	9	10	11
LENGTH OF RAFTER												
LENGTH OF HIP												

RUN OF RAFTER ft.	1	2	3	4	5	6	7	8	9	10
LENGTH OF RAFTER	$1{\cdot}0\frac{1}{4}$	$2{\cdot}0\frac{1}{2}$	$3{\cdot}0\frac{3}{4}$	$4{\cdot}1$	$5{\cdot}1\frac{3}{8}$	$6{\cdot}1\frac{5}{8}$	$7{\cdot}1\frac{7}{8}$	$8{\cdot}2\frac{1}{8}$	$9{\cdot}2\frac{3}{8}$	$10{\cdot}2\frac{5}{8}$
LENGTH OF HIP										

13° PITCH

RISE OF COMMON RAFTER $2\frac{3}{4}''$ PER FOOT OF RUN

BEVELS:

1. COMMON RAFTER	–	SEAT	13
2. 〃 〃	–	RIDGE	77
3. HIP OR VALLEY	–	SEAT	
4. 〃 〃 〃	–	RIDGE	
5. JACK RAFTER	–	EDGE	
6. PURLIN	–	EDGE	
7. 〃	–	SIDE	

JACK RAFTERS 16 in. CENTRES DECREASE

RUN OF RAFTER *ins.*	$\frac{1}{2}$	1	2	3	4	5	6	7	8	9	10	11
LENGTH OF RAFTER												
LENGTH OF HIP												

RUN OF RAFTER *ft.*	1	2	3	4	5	6	7	8	9	10
LENGTH OF RAFTER	$1{\cdot}0\frac{3}{8}$	$2{\cdot}0\frac{5}{8}$	$3{\cdot}0\frac{7}{8}$	$4{\cdot}1\frac{1}{4}$	$5{\cdot}1\frac{5}{8}$	$6{\cdot}1\frac{7}{8}$	$7{\cdot}2\frac{1}{8}$	$8{\cdot}2\frac{1}{2}$	$9{\cdot}2\frac{7}{8}$	$10{\cdot}3\frac{1}{8}$
LENGTH OF HIP										

14° PITCH

RISE OF COMMON RAFTER 3″ PER FOOT OF RUN

BEVELS:

1.	COMMON RAFTER	– SEAT	14
2.	″ ″	– RIDGE	76
3.	HIP OR VALLEY	– SEAT	
4.	″ ″ ″	– RIDGE	
5.	JACK RAFTER	– EDGE	
6.	PURLIN	– EDGE	
7.	″	– SIDE	

JACK RAFTERS 16 in. CENTRES DECREASE

RUN OF RAFTER *ins.*	½	1	2	3	4	5	6	7	8	9	10	11
LENGTH OF RAFTER												
LENGTH OF HIP												

RUN OF RAFTER *ft.*	1	2	3	4	5	6	7	8	9	10
LENGTH OF RAFTER	1·0⅜	2·0¾	3·1⅛	4·1½	5·1⅞	6·2¼	7·2⅝	8·3	9·3⅜	10·3⅝
LENGTH OF HIP										

15° PITCH

RISE OF COMMON RAFTER $3\frac{3}{16}$″ PER FOOT OF RUN

BEVELS: 1. COMMON RAFTER – SEAT 15
2. ″ ″ – RIDGE 75
3. HIP OR VALLEY – SEAT
4. ″ ″ ″ – RIDGE
5. JACK RAFTER – EDGE
6. PURLIN – EDGE
7. ″ – SIDE

JACK RAFTERS 16 in. CENTRES DECREASE

RUN OF RAFTER *ins.*	$\frac{1}{2}$	1	2	3	4	5	6	7	8	9	10	11
LENGTH OF RAFTER												
LENGTH OF HIP												

RUN OF RAFTER *ft.*	1	2	3	4	5	6	7	8	9	10
LENGTH OF RAFTER	$1{\cdot}0\frac{3}{8}$	$2{\cdot}0\frac{7}{8}$	$3{\cdot}1\frac{1}{4}$	$4{\cdot}1\frac{5}{8}$	$5{\cdot}2\frac{1}{8}$	$6{\cdot}2\frac{1}{2}$	7·3	$8{\cdot}3\frac{3}{8}$	$9{\cdot}3\frac{3}{4}$	$10{\cdot}4\frac{1}{4}$
LENGTH OF HIP										

16° or GRECIAN PITCH

RISE OF COMMON RAFTER $3\frac{7}{16}''$ PER FOOT OF RUN

BEVELS:				
1.	COMMON RAFTER	–	SEAT	16
2.	″ ″	–	RIDGE	74
3.	HIP OR VALLEY	–	SEAT	$11\frac{1}{2}$
4.	″ ″ ″	–	RIDGE	$78\frac{1}{2}$
5.	JACK RAFTER	–	EDGE	44
6.	PURLIN	–	EDGE	46
7.	″	–	SIDE	$74\frac{1}{2}$

JACK RAFTERS 16 in. CENTRES DECREASE $16\frac{5}{8}''$, 18 in.—$18\frac{3}{4}''$, 24 in.—2′1″

RUN OF RAFTER *ins.*	$\frac{1}{2}$	1	2	3	4	5	6	7	8	9	10	11
LENGTH OF RAFTER	$\frac{1}{2}$	1	$2\frac{1}{8}$	$3\frac{1}{8}$	$4\frac{1}{8}$	$5\frac{1}{4}$	$6\frac{1}{4}$	$7\frac{1}{4}$	$8\frac{3}{8}$	$9\frac{3}{8}$	$10\frac{3}{8}$	$11\frac{1}{2}$
LENGTH OF HIP	$\frac{3}{4}$	$1\frac{1}{2}$	$2\frac{7}{8}$	$4\frac{3}{8}$	$5\frac{3}{4}$	$7\frac{1}{4}$	$8\frac{5}{8}$	$10\frac{1}{8}$	$11\frac{1}{2}$	13	$14\frac{3}{8}$	$15\frac{7}{8}$

RUN OF RAFTER *ft.*	1	2	3	4	5	6	7	8	9	10
LENGTH OF RAFTER	$1{\cdot}0\frac{1}{2}$	2·1	$3{\cdot}1\frac{1}{2}$	$4{\cdot}1\frac{7}{8}$	$5{\cdot}2\frac{3}{8}$	$6{\cdot}2\frac{7}{8}$	$7{\cdot}3\frac{3}{8}$	$8{\cdot}3\frac{7}{8}$	$9{\cdot}4\frac{7}{8}$	$10{\cdot}4\frac{7}{8}$
LENGTH OF HIP	$1{\cdot}5\frac{1}{4}$	$2{\cdot}10\frac{5}{8}$	4·4	$5{\cdot}9\frac{1}{4}$	$7{\cdot}2\frac{1}{2}$	$8{\cdot}7\frac{7}{8}$	$10{\cdot}1\frac{1}{8}$	$11{\cdot}6\frac{1}{2}$	$12{\cdot}11\frac{3}{4}$	$14{\cdot}5\frac{1}{8}$

17° PITCH

RISE OF COMMON RAFTER $3\frac{11}{16}''$ PER FOOT OF RUN

BEVELS:				
1.	COMMON RAFTER	–	SEAT	17
2.	" "	–	RIDGE	73
3.	HIP OR VALLEY	–	SEAT	12
4.	" " "	–	RIDGE	78
5.	JACK RAFTER	–	EDGE	$43\frac{1}{2}$
6.	PURLIN	–	EDGE	$46\frac{1}{2}$
7.	"	–	SIDE	$73\frac{1}{2}$

JACK RAFTERS 16 in. CENTRES DECREASE $16\frac{3}{4}''$, 18 in.—$18\frac{7}{8}''$, 24 in.—2′ $1\frac{1}{8}''$

RUN OF RAFTER *ins.*	$\frac{1}{2}$	1	2	3	4	5	6	7	8	9	10	11
LENGTH OF RAFTER	$\frac{1}{2}$	1	$2\frac{1}{8}$	$3\frac{1}{8}$	$4\frac{1}{4}$	$5\frac{1}{4}$	$6\frac{1}{4}$	$7\frac{3}{8}$	$8\frac{3}{8}$	$9\frac{3}{8}$	$10\frac{1}{2}$	$11\frac{1}{2}$
LENGTH OF HIP	$\frac{3}{4}$	$1\frac{1}{2}$	$2\frac{7}{8}$	$4\frac{3}{8}$	$5\frac{3}{4}$	$7\frac{1}{4}$	$8\frac{5}{8}$	$10\frac{1}{8}$	$11\frac{1}{2}$	13	$14\frac{1}{2}$	$15\frac{7}{8}$

RUN OF RAFTER *ft.*	1	2	3	4	5	6	7	8	9	10
LENGTH OF RAFTER	$1{\cdot}0\frac{1}{2}$	$2{\cdot}1\frac{1}{8}$	$3{\cdot}1\frac{5}{8}$	$4{\cdot}2\frac{1}{4}$	$5{\cdot}2\frac{3}{4}$	$6{\cdot}3\frac{1}{4}$	$7{\cdot}3\frac{7}{8}$	$8{\cdot}4\frac{3}{8}$	$9{\cdot}4\frac{7}{8}$	$10{\cdot}5\frac{1}{2}$
LENGTH OF HIP	$1{\cdot}5\frac{3}{8}$	$2{\cdot}10\frac{3}{4}$	$4{\cdot}4\frac{1}{8}$	$5{\cdot}9\frac{3}{8}$	$7{\cdot}2\frac{3}{4}$	$8{\cdot}8\frac{1}{8}$	$10{\cdot}1\frac{1}{2}$	$11{\cdot}6\frac{7}{8}$	$13{\cdot}0\frac{1}{4}$	$14{\cdot}5\frac{1}{2}$

$17\frac{1}{2}°$ PITCH

RISE OF COMMON RAFTER $3\frac{3}{4}''$ PER FOOT OF RUN

BEVELS:

1.	COMMON RAFTER	– SEAT	$17\frac{1}{2}$
2.	″ ″	– RIDGE	$72\frac{1}{2}$
3.	HIP OR VALLEY	– SEAT	$12\frac{1}{2}$
4.	″ ″ ″	– RIDGE	$77\frac{1}{2}$
5.	JACK RAFTER	– EDGE	$43\frac{1}{2}$
6.	PURLIN	– EDGE	$46\frac{1}{2}$
7.	″	– SIDE	$73\frac{1}{2}$

JACK RAFTERS 16 in. CENTRES DECREASE $16\frac{3}{4}''$, 18 in.—$18\frac{7}{8}''$, 24 in.—$2'\ 1\frac{1}{8}''$

RUN OF RAFTER *ins.*	$\frac{1}{2}$	1	2	3	4	5	6	7	8	9	10	11
LENGTH OF RAFTER	$\frac{1}{2}$	1	$2\frac{1}{8}$	$3\frac{1}{8}$	$4\frac{1}{4}$	$5\frac{1}{4}$	$6\frac{1}{4}$	$7\frac{3}{8}$	$8\frac{3}{8}$	$9\frac{3}{8}$	$10\frac{1}{2}$	$11\frac{1}{2}$
LENGTH OF HIP	$\frac{3}{4}$	$1\frac{1}{2}$	$2\frac{7}{8}$	$4\frac{3}{8}$	$5\frac{3}{4}$	$7\frac{1}{4}$	$8\frac{5}{8}$	$10\frac{1}{8}$	$11\frac{1}{2}$	13	$14\frac{1}{2}$	$15\frac{7}{8}$

RUN OF RAFTER *ft.*	1	2	3	4	5	6	7	8	9	10
LENGTH OF RAFTER	$1{\cdot}0\frac{5}{8}$	$2{\cdot}1\frac{1}{8}$	$3{\cdot}1\frac{3}{4}$	$4{\cdot}2\frac{3}{8}$	5·3	$6{\cdot}3\frac{1}{2}$	7·4	$8{\cdot}4\frac{5}{8}$	$9{\cdot}5\frac{1}{4}$	$10{\cdot}5\frac{7}{8}$
LENGTH OF HIP	$1{\cdot}5\frac{3}{8}$	$2{\cdot}10\frac{7}{8}$	$4{\cdot}4\frac{1}{4}$	$5{\cdot}9\frac{1}{2}$	7·3	$8{\cdot}8\frac{1}{4}$	$10{\cdot}1\frac{5}{8}$	11·7	$13{\cdot}0\frac{1}{2}$	14·6

18° PITCH

RISE OF COMMON RAFTER $3\frac{7}{8}''$ PER FOOT OF RUN

BEVELS:				
1.	COMMON RAFTER	–	SEAT	18
2.	" "	–	RIDGE	72
3.	HIP OR VALLEY	–	SEAT	13
4.	" " "	–	RIDGE	77
5.	JACK RAFTER	–	EDGE	$43\frac{1}{2}$
6.	PURLIN	–	EDGE	$46\frac{1}{2}$
7.	"	–	SIDE	73

JACK RAFTERS 16 in. CENTRES DECREASE $16\frac{7}{8}''$, 18 in.—$18\frac{7}{8}''$, 24 in.—$2'\ 1\frac{1}{4}''$

RUN OF RAFTER *ins.*	$\frac{1}{2}$	1	2	3	4	5	6	7	8	9	10	11
LENGTH OF RAFTER	$\frac{1}{2}$	1	$2\frac{1}{8}$	$3\frac{1}{8}$	$4\frac{1}{4}$	$5\frac{1}{4}$	$6\frac{3}{8}$	$7\frac{3}{8}$	$8\frac{3}{8}$	$9\frac{1}{2}$	$10\frac{1}{2}$	$11\frac{5}{8}$
LENGTH OF HIP	$\frac{3}{4}$	$1\frac{1}{2}$	$2\frac{7}{8}$	$4\frac{3}{8}$	$5\frac{3}{4}$	$7\frac{1}{4}$	$8\frac{3}{4}$	$10\frac{1}{8}$	$11\frac{5}{8}$	$13\frac{1}{8}$	$14\frac{1}{2}$	16

RUN OF RAFTER *ft.*	1	2	3	4	5	6	7	8	9	10
LENGTH OF RAFTER	$1{\cdot}0\frac{5}{8}$	$2{\cdot}1\frac{1}{4}$	$3{\cdot}1\frac{7}{8}$	$4{\cdot}2\frac{1}{2}$	$5{\cdot}3\frac{1}{8}$	$6{\cdot}3\frac{3}{4}$	$7{\cdot}4\frac{3}{8}$	8·5	$9{\cdot}5\frac{1}{2}$	$10{\cdot}6\frac{1}{8}$
LENGTH OF HIP	$1{\cdot}5\frac{3}{8}$	$2{\cdot}10\frac{7}{8}$	$4{\cdot}4\frac{1}{4}$	$5{\cdot}9\frac{5}{8}$	$7{\cdot}3\frac{1}{8}$	$8{\cdot}8\frac{1}{2}$	10·2	$11{\cdot}7\frac{3}{8}$	$13{\cdot}0\frac{3}{4}$	$14{\cdot}6\frac{1}{4}$

19° PITCH

RISE OF COMMON RAFTER $4\frac{1}{8}''$ PER FOOT OF RUN

BEVELS:
1. COMMON RAFTER – SEAT 19
2. " " – RIDGE 71
3. HIP OR VALLEY – SEAT $13\frac{1}{2}$
4. " " " – RIDGE $76\frac{1}{2}$
5. JACK RAFTER – EDGE $43\frac{1}{2}$
6. PURLIN – EDGE $46\frac{1}{2}$
7. " – SIDE 72

JACK RAFTERS 16 in. CENTRES DECREASE $16\frac{7}{8}''$, 18 in.—19″, 24 in.—2′ $1\frac{3}{8}''$

RUN OF RAFTER *ins.*	$\frac{1}{2}$	1	2	3	4	5	6	7	8	9	10	11
LENGTH OF RAFTER	$\frac{1}{2}$	1	$2\frac{1}{8}$	$3\frac{1}{8}$	$4\frac{1}{4}$	$5\frac{1}{4}$	$6\frac{3}{8}$	$7\frac{3}{8}$	$8\frac{1}{2}$	$9\frac{1}{2}$	$10\frac{5}{8}$	$11\frac{5}{8}$
LENGTH OF HIP	$\frac{3}{4}$	$1\frac{1}{2}$	$2\frac{7}{8}$	$4\frac{3}{8}$	$5\frac{7}{8}$	$7\frac{1}{4}$	$8\frac{3}{4}$	$10\frac{1}{8}$	$11\frac{5}{8}$	$13\frac{1}{8}$	$14\frac{1}{2}$	16

RUN OF RAFTER *ft.*	1	2	3	4	5	6	7	8	9	10
LENGTH OF RAFTER	$1{\cdot}0\frac{3}{4}$	$2{\cdot}1\frac{3}{8}$	$3{\cdot}2\frac{1}{8}$	$4{\cdot}2\frac{3}{4}$	$5{\cdot}3\frac{1}{2}$	$6{\cdot}4\frac{1}{8}$	$7{\cdot}4\frac{7}{8}$	$8{\cdot}5\frac{1}{2}$	$9{\cdot}6\frac{1}{4}$	$10{\cdot}6\frac{7}{8}$
LENGTH OF HIP	$1{\cdot}5\frac{1}{2}$	$2{\cdot}10\frac{7}{8}$	$4{\cdot}4\frac{3}{8}$	$5{\cdot}9\frac{7}{8}$	$7{\cdot}3\frac{1}{4}$	$8{\cdot}8\frac{3}{4}$	$10{\cdot}2\frac{1}{4}$	$11{\cdot}7\frac{5}{8}$	$13{\cdot}1\frac{1}{8}$	$14{\cdot}6\frac{1}{2}$

20° PITCH

RISE OF COMMON RAFTER $4\frac{3}{8}''$ PER FOOT OF RUN

BEVELS:				
1.	COMMON RAFTER	–	SEAT	20
2.	″ ″	–	RIDGE	70
3.	HIP OR VALLEY	–	SEAT	$14\frac{1}{2}$
4.	″ ″ ″	–	RIDGE	$75\frac{1}{2}$
5.	JACK RAFTER	–	EDGE	43
6.	PURLIN	–	EDGE	47
7.	″	–	SIDE	71

JACK RAFTERS 16 in. CENTRES DECREASE 17″, 18 in.—$19\frac{1}{8}''$, 24 in.—2′ $1\frac{1}{2}''$

RUN OF RAFTER *ins.*	$\frac{1}{2}$	1	2	3	4	5	6	7	8	9	10	11
LENGTH OF RAFTER	$\frac{1}{2}$	$1\frac{1}{8}$	$2\frac{1}{8}$	$3\frac{1}{4}$	$4\frac{1}{4}$	$5\frac{3}{8}$	$6\frac{3}{8}$	$7\frac{1}{2}$	$8\frac{1}{2}$	$9\frac{5}{8}$	$10\frac{5}{8}$	$11\frac{3}{4}$
LENGTH OF HIP	$\frac{3}{4}$	$1\frac{1}{2}$	$2\frac{7}{8}$	$4\frac{3}{8}$	$5\frac{7}{8}$	$7\frac{1}{4}$	$8\frac{3}{4}$	$10\frac{1}{4}$	$11\frac{5}{8}$	$13\frac{1}{8}$	$14\frac{5}{8}$	16

RUN OF RAFTER *ft.*	1	2	3	4	5	6	7	8	9	10
LENGTH OF RAFTER	$1{\cdot}0\frac{3}{4}$	$2{\cdot}1\frac{1}{2}$	$3{\cdot}2\frac{1}{4}$	$4{\cdot}3\frac{1}{8}$	$5{\cdot}3\frac{7}{8}$	$6{\cdot}4\frac{5}{8}$	$7{\cdot}5\frac{3}{8}$	$8{\cdot}6\frac{1}{8}$	$9{\cdot}6\frac{7}{8}$	$10{\cdot}7\frac{3}{4}$
LENGTH OF HIP	$1{\cdot}5\frac{1}{2}$	2·11	$4{\cdot}4\frac{1}{2}$	$5{\cdot}10\frac{1}{8}$	$7{\cdot}3\frac{5}{8}$	$8{\cdot}9\frac{1}{8}$	$10{\cdot}2\frac{5}{8}$	$11{\cdot}8\frac{1}{8}$	$13{\cdot}1\frac{5}{8}$	$14{\cdot}7\frac{1}{8}$

21° PITCH

RISE OF COMMON RAFTER $4\frac{5}{8}''$ PER FOOT OF RUN

BEVELS:				
1.	COMMON RAFTER	–	SEAT	21
2.	" "	–	RIDGE	69
3.	HIP OR VALLEY	–	SEAT	15
4.	" " "	–	RIDGE	75
5.	JACK RAFTER	–	EDGE	43
6.	PURLIN	–	EDGE	47
7.	"	–	SIDE	$70\frac{1}{2}$

JACK RAFTERS 16 in. CENTRES DECREASE $17\frac{1}{8}''$, 18 in.—$19\frac{1}{4}''$, 24 in.—$2'\ 1\frac{3}{4}''$

RUN OF RAFTER ins.	$\frac{1}{2}$	1	2	3	4	5	6	7	8	9	10	11
LENGTH OF RAFTER	$\frac{1}{2}$	$1\frac{1}{8}$	$2\frac{1}{8}$	$3\frac{1}{4}$	$4\frac{1}{4}$	$5\frac{3}{8}$	$6\frac{3}{8}$	$7\frac{1}{2}$	$8\frac{5}{8}$	$9\frac{5}{8}$	$10\frac{3}{4}$	$11\frac{3}{4}$
LENGTH OF HIP	$\frac{3}{4}$	$1\frac{1}{2}$	$2\frac{7}{8}$	$4\frac{3}{8}$	$5\frac{7}{8}$	$7\frac{3}{8}$	$8\frac{3}{4}$	$10\frac{1}{4}$	$11\frac{3}{4}$	$13\frac{1}{8}$	$14\frac{5}{8}$	$16\frac{1}{8}$

RUN OF RAFTER ft.	1	2	3	4	5	6	7	8	9	10
LENGTH OF RAFTER	$1{\cdot}0\frac{7}{8}$	$2{\cdot}1\frac{3}{4}$	$3{\cdot}2\frac{1}{2}$	$4{\cdot}3\frac{3}{8}$	$5{\cdot}4\frac{1}{4}$	$6{\cdot}5\frac{1}{8}$	7·6	$8{\cdot}6\frac{7}{8}$	$9{\cdot}7\frac{5}{8}$	$10{\cdot}8\frac{1}{2}$
LENGTH OF HIP	$1{\cdot}5\frac{3}{8}$	$2{\cdot}11\frac{1}{8}$	$4{\cdot}4\frac{3}{4}$	$5{\cdot}10\frac{1}{4}$	$7{\cdot}3\frac{7}{8}$	$8{\cdot}9\frac{1}{2}$	10·3	$11{\cdot}8\frac{5}{8}$	$13{\cdot}2\frac{1}{8}$	$14{\cdot}7\frac{3}{4}$

22° PITCH

RISE OF COMMON RAFTER $4\frac{7}{8}''$ PER FOOT OF RUN

BEVELS:				
1.	COMMON RAFTER	–	SEAT	22
2.	″ ″	–	RIDGE	68
3.	HIP OR VALLEY	–	SEAT	16
4.	″ ″ ″	–	RIDGE	74
5.	JACK RAFTER	–	EDGE	43
6.	PURLIN	–	EDGE	47
7.	″	–	SIDE	$69\frac{1}{2}$

JACK RAFTERS 16 in. CENTRES DECREASE $17\frac{1}{4}''$, 18 in.—$19\frac{3}{8}''$, 24 in.—2′ $1\frac{7}{8}''$

RUN OF RAFTER ins.	$\frac{1}{2}$	1	2	3	4	5	6	7	8	9	10	11
LENGTH OF RAFTER	$\frac{1}{2}$	$1\frac{1}{8}$	$2\frac{1}{8}$	$3\frac{1}{4}$	$4\frac{3}{8}$	$5\frac{3}{8}$	$6\frac{1}{2}$	$7\frac{1}{2}$	$8\frac{5}{8}$	$9\frac{3}{4}$	$10\frac{3}{4}$	$11\frac{7}{8}$
LENGTH OF HIP	$\frac{3}{4}$	$1\frac{1}{2}$	3	$4\frac{3}{8}$	$5\frac{7}{8}$	$7\frac{3}{8}$	$8\frac{7}{8}$	$10\frac{1}{4}$	$11\frac{3}{4}$	$13\frac{1}{4}$	$14\frac{3}{4}$	$16\frac{1}{8}$

RUN OF RAFTER ft.	1	2	3	4	5	6	7	8	9	10
LENGTH OF RAFTER	1·1	$2{\cdot}1\frac{7}{8}$	$3{\cdot}2\frac{7}{8}$	$4{\cdot}3\frac{3}{4}$	$5{\cdot}4\frac{3}{4}$	$6{\cdot}5\frac{5}{8}$	$7{\cdot}6\frac{5}{8}$	$8{\cdot}7\frac{1}{2}$	$9{\cdot}8\frac{1}{2}$	$10{\cdot}9\frac{3}{8}$
LENGTH OF HIP	$1{\cdot}5\frac{5}{8}$	$2{\cdot}11\frac{1}{4}$	$4{\cdot}4\frac{7}{8}$	$5{\cdot}10\frac{5}{8}$	$7{\cdot}4\frac{1}{4}$	$8{\cdot}9\frac{7}{8}$	$10{\cdot}3\frac{1}{2}$	$11{\cdot}9\frac{1}{8}$	$13{\cdot}2\frac{7}{8}$	$14{\cdot}8\frac{1}{2}$

$22\frac{1}{2}^{\circ}$ PITCH

RISE OF COMMON RAFTER 5″ PER FOOT OF RUN

BEVELS:

1.	COMMON RAFTER	–	SEAT	$22\frac{1}{2}$
2.	″ ″	–	RIDGE	$67\frac{1}{2}$
3.	HIP OR VALLEY	–	SEAT	$16\frac{1}{4}$
4.	″ ″ ″	–	RIDGE	$73\frac{3}{4}$
5.	JACK RAFTER	–	EDGE	$42\frac{3}{4}$
6.	PURLIN	–	EDGE	$47\frac{1}{2}$
7.	″	–	SIDE	69

JACK RAFTERS 16 in. CENTRES DECREASE $17\frac{1}{4}$″, 18 in.—$19\frac{3}{8}$″, 24 in.—2′ 2″

RUN OF RAFTER *ins.*	$\frac{1}{2}$	1	2	3	4	5	6	7	8	9	10	11
LENGTH OF RAFTER	$\frac{1}{2}$	$1\frac{1}{8}$	$2\frac{1}{8}$	$3\frac{1}{4}$	$4\frac{3}{8}$	$5\frac{3}{8}$	$6\frac{1}{2}$	$7\frac{1}{2}$	$8\frac{5}{8}$	$9\frac{3}{4}$	$10\frac{3}{4}$	$11\frac{7}{8}$
LENGTH OF HIP	$\frac{3}{4}$	$1\frac{1}{2}$	3	$4\frac{3}{8}$	$5\frac{7}{8}$	$7\frac{3}{8}$	$8\frac{7}{8}$	$10\frac{1}{4}$	$11\frac{3}{4}$	$13\frac{1}{4}$	$14\frac{3}{4}$	$16\frac{1}{8}$

RUN OF RAFTER *ft.*	1	2	3	4	5	6	7	8	9	10
LENGTH OF RAFTER	1·1	2·2	3·3	4·4	5·5	6·6	7·7	8·8	$9{\cdot}8\frac{7}{8}$	$10{\cdot}9\frac{7}{8}$
LENGTH OF HIP	$1{\cdot}5\frac{5}{8}$	$2{\cdot}11\frac{1}{4}$	4·5	$5{\cdot}10\frac{3}{4}$	$7{\cdot}5\frac{7}{8}$	$8{\cdot}9\frac{1}{2}$	$10{\cdot}3\frac{3}{4}$	$11{\cdot}9\frac{1}{2}$	13·3	$14{\cdot}8\frac{3}{4}$

23° PITCH

RISE OF COMMON RAFTER 5$\frac{1}{8}$″ PER FOOT OF RUN

BEVELS:

1.	COMMON RAFTER	–	SEAT	23
2.	″ ″	–	RIDGE	67
3.	HIP OR VALLEY	–	SEAT	16$\frac{1}{2}$
4.	″ ″ ″	–	RIDGE	73$\frac{1}{2}$
5.	JACK RAFTER	–	EDGE	42$\frac{1}{2}$
6.	PURLIN	–	EDGE	47$\frac{1}{2}$
7.	″	–	SIDE	68$\frac{1}{2}$

JACK RAFTERS 16 in. CENTRES DECREASE 17$\frac{3}{8}$″, 18 in.—19$\frac{1}{2}$″, 24 in.—2′ 2$\frac{1}{8}$″

RUN OF RAFTER *ins.*	$\frac{1}{2}$	1	2	3	4	5	6	7	8	9	10	11
LENGTH OF RAFTER	$\frac{1}{2}$	1$\frac{1}{8}$	2$\frac{1}{8}$	3$\frac{1}{4}$	4$\frac{3}{8}$	5$\frac{3}{8}$	6$\frac{1}{2}$	7$\frac{5}{8}$	8$\frac{3}{4}$	9$\frac{3}{4}$	10$\frac{7}{8}$	12
LENGTH OF HIP	$\frac{3}{4}$	1$\frac{1}{2}$	3	4$\frac{3}{8}$	5$\frac{7}{8}$	7$\frac{3}{8}$	8$\frac{7}{8}$	10$\frac{3}{8}$	11$\frac{3}{4}$	13$\frac{1}{4}$	14$\frac{3}{4}$	16$\frac{1}{4}$

RUN OF RAFTER *ft.*	1	2	3	4	5	6	7	8	9	10
LENGTH OF RAFTER	1·1	2·2$\frac{1}{8}$	3·3$\frac{1}{8}$	4·4$\frac{1}{8}$	5·5$\frac{1}{8}$	6·6$\frac{1}{4}$	7·7$\frac{1}{4}$	8·8$\frac{1}{4}$	9·9$\frac{3}{8}$	10·10$\frac{3}{8}$
LENGTH OF HIP	1·5$\frac{3}{4}$	2·11$\frac{3}{8}$	4·5$\frac{1}{8}$	5·10$\frac{7}{8}$	7·4$\frac{1}{2}$	8·10$\frac{1}{4}$	10·4	11·9$\frac{3}{4}$	13·3$\frac{1}{2}$	14·9$\frac{1}{8}$

24° or ROMAN PITCH

RISE OF COMMON RAFTER $5\frac{5}{16}''$ PER FOOT OF RUN

BEVELS:

1.	COMMON RAFTER	–	SEAT	24
2.	" "	–	RIDGE	66
3.	HIP OR VALLEY	–	SEAT	$17\frac{1}{2}$
4.	" " "	–	RIDGE	$72\frac{1}{2}$
5.	JACK RAFTER	–	EDGE	$42\frac{1}{2}$
6.	PURLIN	–	EDGE	$47\frac{1}{2}$
7.	"	–	SIDE	68

JACK RAFTERS 16 in. CENTRES DECREASE $17\frac{1}{2}''$, 18 in.—$19\frac{3}{4}''$, 24 in.—$2'\ 2\frac{1}{4}''$

RUN OF RAFTER *ins.*	$\frac{1}{2}$	1	2	3	4	5	6	7	8	9	10	11
LENGTH OF RAFTER	$\frac{1}{2}$	$1\frac{1}{8}$	$2\frac{1}{8}$	$3\frac{1}{4}$	$4\frac{3}{8}$	$5\frac{1}{2}$	$6\frac{1}{2}$	$7\frac{5}{8}$	$8\frac{3}{4}$	$9\frac{7}{8}$	11	12
LENGTH OF HIP	$\frac{3}{4}$	$1\frac{1}{2}$	3	$4\frac{1}{2}$	$5\frac{7}{8}$	$7\frac{3}{8}$	$8\frac{7}{8}$	$10\frac{3}{8}$	$11\frac{7}{8}$	$13\frac{3}{8}$	$14\frac{7}{8}$	$16\frac{1}{4}$

RUN OF RAFTER *ft.*	1	2	3	4	5	6	7	8	9	10
LENGTH OF RAFTER	$1{\cdot}1\frac{1}{8}$	$2{\cdot}2\frac{1}{4}$	$3{\cdot}3\frac{3}{8}$	$4{\cdot}4\frac{1}{2}$	$5{\cdot}5\frac{5}{8}$	$6{\cdot}6\frac{3}{4}$	7·8	$8{\cdot}9\frac{1}{8}$	$9{\cdot}10\frac{1}{4}$	$10{\cdot}11\frac{3}{8}$
LENGTH OF HIP	$1{\cdot}5\frac{3}{4}$	$2{\cdot}11\frac{5}{8}$	$4{\cdot}5\frac{3}{8}$	$5{\cdot}11\frac{1}{8}$	7·5	$8{\cdot}10\frac{3}{4}$	$10{\cdot}4\frac{5}{8}$	$11{\cdot}10\frac{3}{8}$	$13{\cdot}4\frac{1}{8}$	14·10

25°

RISE OF COMMON RAFTER $5\frac{9}{16}''$ PER FOOT OF RUN

BEVELS:				
1.	COMMON RAFTER	–	SEAT	25
2.	" "	–	RIDGE	65
3.	HIP OR VALLEY	–	SEAT	17
4.	" " "	–	RIDGE	72
5.	JACK RAFTER	–	EDGE	42
6.	PURLIN	–	EDGE	48
7.	"	–	SIDE	67

JACK RAFTERS 16 in. CENTRES DECREASE $17\frac{5}{8}''$, 18 in.—$19\frac{7}{8}''$, 24 in.—$2'\ 2\frac{1}{2}''$

RUN OF RAFTER *ins.*	$\frac{1}{2}$	1	2	3	4	5	6	7	8	9	10	11
LENGTH OF RAFTER	$\frac{1}{2}$	$1\frac{1}{8}$	$2\frac{1}{4}$	$3\frac{3}{8}$	$4\frac{3}{8}$	$5\frac{1}{2}$	$6\frac{5}{8}$	$7\frac{3}{4}$	$8\frac{7}{8}$	$9\frac{7}{8}$	11	$12\frac{1}{8}$
LENGTH OF HIP	$\frac{3}{4}$	$1\frac{1}{2}$	3	$4\frac{1}{2}$	6	$7\frac{1}{2}$	$8\frac{7}{8}$	$10\frac{3}{8}$	$11\frac{7}{8}$	$13\frac{3}{8}$	$14\frac{7}{8}$	$16\frac{3}{8}$

RUN OF RAFTER *ft.*	1	2	3	4	5	6	7	8	9	10
LENGTH OF RAFTER	$1{\cdot}1\frac{1}{4}$	$2{\cdot}2\frac{1}{2}$	$3{\cdot}3\frac{3}{4}$	4·5	$5{\cdot}6\frac{1}{4}$	$6{\cdot}7\frac{1}{2}$	$7{\cdot}8\frac{5}{8}$	$8{\cdot}9\frac{7}{8}$	$9{\cdot}11\frac{1}{8}$	$11{\cdot}0\frac{3}{8}$
LENGTH OF HIP	$1{\cdot}5\frac{7}{8}$	$2{\cdot}11\frac{3}{4}$	$4{\cdot}5\frac{5}{8}$	$5{\cdot}11\frac{1}{2}$	$7{\cdot}5\frac{3}{8}$	$8{\cdot}11\frac{1}{4}$	$10{\cdot}5\frac{1}{8}$	11·11	$13{\cdot}4\frac{3}{4}$	$14{\cdot}10\frac{5}{8}$

Note: Quarter Pitch (Rise = $\frac{1}{4}$ Span) has a rafter seat bevel of 26° 34′. Lengths of rafters, etc., are based on this angle

QUARTER PITCH

RISE OF COMMON RAFTER 6″ PER FOOT OF RUN

BEVELS:

1. COMMON RAFTER	–	SEAT	$26\frac{1}{2}$
2. 〃 〃	–	RIDGE	$63\frac{1}{2}$
3. HIP OR VALLEY	–	SEAT	$19\frac{1}{2}$
4. 〃 〃 〃	–	RIDGE	$70\frac{1}{2}$
5. JACK RAFTER	–	EDGE	42
6. PURLIN	–	EDGE	48
7. 〃	–	SIDE	66

JACK RAFTERS 16 in. CENTRES DECREASE $17\frac{7}{8}$″, 18 in.—$20\frac{1}{8}$″, 24 in.—2′ $2\frac{7}{8}$″

RUN OF RAFTER *ins.*	$\frac{1}{2}$	1	2	3	4	5	6	7	8	9	10	11
LENGTH OF RAFTER	$\frac{1}{2}$	$1\frac{1}{8}$	$2\frac{1}{4}$	$3\frac{3}{8}$	$4\frac{1}{2}$	$5\frac{5}{8}$	$6\frac{3}{4}$	$7\frac{7}{8}$	9	10	$11\frac{1}{4}$	$12\frac{1}{4}$
LENGTH OF HIP	$\frac{3}{4}$	$1\frac{1}{2}$	3	$4\frac{1}{2}$	6	$7\frac{1}{2}$	9	$10\frac{1}{2}$	12	$13\frac{1}{2}$	15	$16\frac{1}{2}$

RUN OF RAFTER *ft.*	1	2	3	4	5	6	7	8	9	10
LENGTH OF RAFTER	$1{\cdot}1\frac{3}{8}$	$2{\cdot}2\frac{7}{8}$	$3{\cdot}4\frac{1}{4}$	$4{\cdot}5\frac{5}{8}$	$5{\cdot}7\frac{1}{8}$	$6{\cdot}8\frac{1}{2}$	$7{\cdot}9\frac{7}{8}$	$8{\cdot}11\frac{3}{8}$	$10{\cdot}0\frac{3}{4}$	$11{\cdot}2\frac{1}{8}$
LENGTH OF HIP	1·6	3·0	4·6	6·0	7·6	9·0	10·6	12·0	13·6	15·0

$27\frac{1}{2}°$ PITCH

RISE OF COMMON RAFTER $6\frac{1}{4}''$ PER FOOT OF RUN

BEVELS:				
1.	COMMON RAFTER	–	SEAT	$27\frac{1}{2}$
2.	" "	–	RIDGE	$62\frac{1}{2}$
3.	HIP OR VALLEY	–	SEAT	20
4.	" " "	–	RIDGE	70
5.	JACK RAFTER	–	EDGE	$41\frac{3}{4}$
6.	PURLIN	–	EDGE	$48\frac{1}{2}$
7.	"	–	SIDE	65

JACK RAFTERS 16 in. CENTRES DECREASE $18\frac{1}{8}''$, 18 in.—$20\frac{3}{8}''$, 24 in.—2′ 3″

RUN OF RAFTER *ins.*	$\frac{1}{2}$	1	2	3	4	5	6	7	8	9	10	11
LENGTH OF RAFTER	$\frac{5}{8}$	$1\frac{1}{8}$	$2\frac{1}{4}$	$3\frac{3}{8}$	$4\frac{1}{2}$	$5\frac{5}{8}$	$6\frac{3}{4}$	$7\frac{7}{8}$	9	$10\frac{1}{8}$	$11\frac{1}{4}$	$12\frac{3}{8}$
LENGTH OF HIP	$\frac{3}{4}$	$1\frac{1}{2}$	3	$4\frac{1}{2}$	6	$7\frac{1}{2}$	$9\frac{1}{8}$	$10\frac{5}{8}$	12	$13\frac{1}{2}$	15	$16\frac{1}{2}$

RUN OF RAFTER *ft.*	1	2	3	4	5	6	7	8	9	10
LENGTH OF RAFTER	$1{\cdot}1\frac{1}{2}$	2·3	$3{\cdot}4\frac{1}{2}$	$4{\cdot}6\frac{1}{8}$	$5{\cdot}7\frac{5}{8}$	$6{\cdot}9\frac{1}{8}$	$7{\cdot}10\frac{5}{8}$	$9{\cdot}0\frac{1}{4}$	$10{\cdot}1\frac{3}{4}$	$11{\cdot}3\frac{1}{4}$
LENGTH OF HIP	1·6	3·0	$4{\cdot}6\frac{1}{8}$	$6{\cdot}0\frac{3}{8}$	$7{\cdot}6\frac{3}{8}$	$9{\cdot}0\frac{1}{2}$	$10{\cdot}6\frac{5}{8}$	$12{\cdot}0\frac{5}{8}$	$13{\cdot}6\frac{3}{4}$	$15{\cdot}0\frac{7}{8}$

28° PITCH

RISE OF COMMON RAFTER $6\frac{3}{8}''$ PER FOOT OF RUN

BEVELS:				
1.	COMMON RAFTER	–	SEAT	28
2.	″ ″	–	RIDGE	62
3.	HIP OR VALLEY	–	SEAT	$20\frac{1}{2}$
4.	″ ″ ″	–	RIDGE	$69\frac{1}{2}$
5.	JACK RAFTER	–	EDGE	$41\frac{1}{2}$
6.	PURLIN	–	EDGE	$48\frac{1}{2}$
7.	″	–	SIDE	65

JACK RAFTERS 16 in. CENTRES DECREASE $18\frac{1}{8}''$, 18 in.—$20\frac{3}{8}''$, 24 in.—$2'\ 3\frac{1}{8}''$

RUN OF RAFTER *ins.*	$\frac{1}{2}$	1	2	3	4	5	6	7	8	9	10	11
LENGTH OF RAFTER	$\frac{5}{8}$	$1\frac{1}{8}$	$2\frac{1}{4}$	$3\frac{3}{8}$	$4\frac{1}{2}$	$5\frac{5}{8}$	$6\frac{3}{4}$	$7\frac{7}{8}$	9	$10\frac{1}{4}$	$11\frac{3}{8}$	$12\frac{1}{2}$
LENGTH OF HIP	$\frac{3}{4}$	$1\frac{1}{2}$	3	$4\frac{1}{2}$	6	$7\frac{1}{2}$	$9\frac{1}{8}$	$10\frac{5}{8}$	$12\frac{1}{8}$	$13\frac{5}{8}$	$15\frac{1}{8}$	$16\frac{5}{8}$

RUN OF RAFTER *ft.*	1	2	3	4	5	6	7	8	9	10
LENGTH OF RAFTER	$1{\cdot}1\frac{5}{8}$	$2{\cdot}3\frac{1}{8}$	$3{\cdot}4\frac{3}{4}$	$4{\cdot}6\frac{3}{8}$	5·8	$6{\cdot}9\frac{1}{2}$	$7{\cdot}11\frac{1}{8}$	$9{\cdot}0\frac{3}{4}$	$10{\cdot}2\frac{3}{8}$	$11{\cdot}3\frac{7}{8}$
LENGTH OF HIP	$1{\cdot}6\frac{1}{8}$	$3{\cdot}0\frac{1}{4}$	$4{\cdot}6\frac{3}{8}$	$6{\cdot}0\frac{1}{2}$	$7{\cdot}6\frac{5}{8}$	$9{\cdot}0\frac{3}{4}$	$10{\cdot}6\frac{7}{8}$	12·1	$13{\cdot}7\frac{1}{8}$	$15{\cdot}1\frac{3}{8}$

29° PITCH

RISE OF COMMON RAFTER $6\frac{5}{8}''$ PER FOOT OF RUN

BEVELS:				
1.	COMMON RAFTER	–	SEAT	29
2.	" "	–	RIDGE	61
3.	HIP OR VALLEY	–	SEAT	$21\frac{1}{2}$
4.	" " "	–	RIDGE	$68\frac{1}{2}$
5.	JACK RAFTER	–	EDGE	41
6.	PURLIN	–	EDGE	49
7.	"	–	SIDE	64

JACK RAFTERS 16 in. CENTRES DECREASE $18\frac{1}{4}''$, 18 in.—$20\frac{5}{8}''$, 24 in.—2′ $3\frac{1}{2}''$

RUN OF RAFTER *ins.*	$\frac{1}{2}$	1	2	3	4	5	6	7	8	9	10	11
LENGTH OF RAFTER	$\frac{5}{8}$	$1\frac{1}{8}$	$2\frac{1}{4}$	$3\frac{3}{8}$	$4\frac{5}{8}$	$5\frac{3}{4}$	$6\frac{7}{8}$	8	$9\frac{1}{8}$	$10\frac{1}{4}$	$11\frac{1}{2}$	$12\frac{5}{8}$
LENGTH OF HIP	$\frac{3}{4}$	$1\frac{1}{2}$	3	$4\frac{5}{8}$	$6\frac{1}{8}$	$7\frac{5}{8}$	$9\frac{1}{8}$	$10\frac{5}{8}$	$12\frac{1}{8}$	$13\frac{5}{8}$	$15\frac{1}{4}$	$16\frac{3}{4}$

RUN OF RAFTER *ft.*	1	2	3	4	5	6	7	8	9	10
LENGTH OF RAFTER	$1{\cdot}1\frac{3}{4}$	$2{\cdot}3\frac{1}{2}$	$3{\cdot}5\frac{1}{8}$	$4{\cdot}6\frac{7}{8}$	$5{\cdot}8\frac{5}{8}$	$6{\cdot}10\frac{3}{8}$	8·0	$9{\cdot}1\frac{3}{4}$	$10{\cdot}3\frac{1}{2}$	$11{\cdot}5\frac{1}{4}$
LENGTH OF HIP	$1{\cdot}6\frac{1}{4}$	$3{\cdot}0\frac{1}{2}$	$4{\cdot}6\frac{5}{8}$	$6{\cdot}0\frac{7}{8}$	$7{\cdot}7\frac{1}{8}$	$9{\cdot}1\frac{3}{8}$	$10{\cdot}7\frac{5}{8}$	$12{\cdot}1\frac{7}{8}$	13·8	$15{\cdot}2\frac{1}{4}$

30° PITCH

RISE OF COMMON RAFTER $6\frac{15}{16}$″ PER FOOT OF RUN

BEVELS:				
1.	COMMON RAFTER	–	SEAT	30
2.	″ ″	–	RIDGE	60
3.	HIP OR VALLEY	–	SEAT	22
4.	″ ″ ″	–	RIDGE	68
5.	JACK RAFTER	–	EDGE	41
6.	PURLIN	–	EDGE	49
7.	″	–	SIDE	$63\frac{1}{2}$

JACK RAFTERS 16 in. CENTRES DECREASE $18\frac{1}{2}$″, 18 in.—$20\frac{3}{4}$″, 24 in.—2′ $3\frac{3}{4}$″

RUN OF RAFTER *ins.*	$\frac{1}{2}$	1	2	3	4	5	6	7	8	9	10	11
LENGTH OF RAFTER	$\frac{5}{8}$	$1\frac{1}{8}$	$2\frac{1}{4}$	$3\frac{1}{2}$	$4\frac{5}{8}$	$5\frac{3}{4}$	$6\frac{7}{8}$	$8\frac{1}{8}$	$9\frac{1}{4}$	$10\frac{3}{8}$	$11\frac{1}{2}$	$12\frac{3}{4}$
LENGTH OF HIP	$\frac{3}{4}$	$1\frac{1}{2}$	3	$4\frac{5}{8}$	$6\frac{1}{8}$	$7\frac{5}{8}$	$9\frac{1}{8}$	$10\frac{3}{4}$	$12\frac{1}{4}$	$13\frac{3}{4}$	$15\frac{1}{4}$	$16\frac{3}{4}$

RUN OF RAFTER *ft.*	1	2	3	4	5	6	7	8	9	10
LENGTH OF RAFTER	$1{\cdot}1\frac{7}{8}$	$2{\cdot}3\frac{3}{4}$	$3{\cdot}5\frac{5}{8}$	$4{\cdot}7\frac{3}{8}$	$5{\cdot}9\frac{1}{4}$	$6{\cdot}11\frac{1}{8}$	8·1	$9{\cdot}2\frac{7}{8}$	$10{\cdot}4\frac{3}{4}$	$11{\cdot}6\frac{5}{8}$
LENGTH OF HIP	$1{\cdot}6\frac{3}{8}$	$3{\cdot}0\frac{5}{8}$	4·7	$6{\cdot}1\frac{1}{4}$	$7{\cdot}7\frac{5}{8}$	9·2	$10{\cdot}8\frac{3}{8}$	$12{\cdot}2\frac{5}{8}$	13·9	$15{\cdot}3\frac{1}{4}$

31° PITCH

RISE OF COMMON RAFTER $7\frac{3}{16}''$ PER FOOT OF RUN

BEVELS:

1.	COMMON RAFTER	–	SEAT	31
2.	″ ″	–	RIDGE	59
3.	HIP OR VALLEY	–	SEAT	23
4.	″ ″ ″	–	RIDGE	67
5.	JACK RAFTER	–	EDGE	$40\frac{1}{2}$
6.	PURLIN	–	EDGE	$49\frac{1}{2}$
7.	″	–	SIDE	$62\frac{1}{2}$

JACK RAFTERS 16 in. CENTRES DECREASE $18\frac{5}{8}''$, 18 in.—21″, 24 in.—2′ 4″

RUN OF RAFTER *ins.*	$\frac{1}{2}$	1	2	3	4	5	6	7	8	9	10	11
LENGTH OF RAFTER	$\frac{5}{8}$	$1\frac{1}{8}$	$2\frac{3}{8}$	$3\frac{1}{2}$	$4\frac{5}{8}$	$5\frac{7}{8}$	7	$8\frac{1}{8}$	$9\frac{3}{8}$	$10\frac{1}{2}$	$11\frac{5}{8}$	$12\frac{7}{8}$
LENGTH OF HIP	$\frac{3}{4}$	$1\frac{1}{2}$	$3\frac{1}{8}$	$4\frac{5}{8}$	$6\frac{1}{8}$	$7\frac{5}{8}$	$9\frac{1}{4}$	$10\frac{3}{4}$	$12\frac{1}{4}$	$13\frac{7}{8}$	$15\frac{3}{8}$	$16\frac{7}{8}$

RUN OF RAFTER *ft.*	1	2	3	4	5	6	7	8	9	10
LENGTH OF RAFTER	1·2	2·4	3·6	4·8	5·10	7·0	8·2	9·4	10·6	11·8
LENGTH OF HIP	$1{\cdot}6\frac{1}{2}$	$3{\cdot}0\frac{7}{8}$	$4{\cdot}7\frac{1}{4}$	$6{\cdot}1\frac{3}{4}$	$7{\cdot}8\frac{1}{8}$	$9{\cdot}2\frac{5}{8}$	10·9	$12{\cdot}3\frac{1}{2}$	13·10	$15{\cdot}4\frac{3}{8}$

32° PITCH

RISE OF COMMON RAFTER $7\frac{1}{2}''$ PER FOOT OF RUN

BEVELS:

1.	COMMON RAFTER	–	SEAT	32
2.	" "	–	RIDGE	58
3.	HIP OR VALLEY	–	SEAT	24
4.	" " "	–	RIDGE	66
5.	JACK RAFTER	–	EDGE	$40\frac{1}{2}$
6.	PURLIN	–	EDGE	$49\frac{1}{2}$
7.	"	–	SIDE	62

JACK RAFTERS 16 in. CENTRES DECREASE $18\frac{7}{8}''$, 18 in.—$21\frac{1}{4}''$, 24 in.—2′ $4\frac{1}{4}''$

RUN OF RAFTER ins.	$\frac{1}{2}$	1	2	3	4	5	6	7	8	9	10	11
LENGTH OF RAFTER	$\frac{5}{8}$	$1\frac{1}{8}$	$2\frac{3}{8}$	$3\frac{1}{2}$	$4\frac{3}{4}$	$5\frac{7}{8}$	$7\frac{1}{8}$	$8\frac{1}{4}$	$9\frac{3}{8}$	$10\frac{5}{8}$	$11\frac{3}{4}$	13
LENGTH OF HIP	$\frac{3}{4}$	$1\frac{1}{2}$	$3\frac{1}{8}$	$4\frac{5}{8}$	$6\frac{1}{8}$	$7\frac{3}{4}$	$9\frac{1}{4}$	$10\frac{7}{8}$	$12\frac{3}{8}$	$13\frac{7}{8}$	$15\frac{1}{2}$	17

RUN OF RAFTER ft.	1	2	3	4	5	6	7	8	9	10
LENGTH OF RAFTER	$1{\cdot}2\frac{1}{8}$	$2{\cdot}4\frac{1}{4}$	$3{\cdot}6\frac{1}{2}$	$4{\cdot}8\frac{5}{8}$	$5{\cdot}10\frac{3}{4}$	$7{\cdot}0\frac{7}{8}$	8·3	$9{\cdot}5\frac{1}{4}$	$10{\cdot}7\frac{3}{8}$	$11{\cdot}9\frac{1}{2}$
LENGTH OF HIP	$1{\cdot}6\frac{1}{2}$	$3{\cdot}1\frac{1}{8}$	$4{\cdot}7\frac{5}{8}$	$6{\cdot}2\frac{1}{4}$	$7{\cdot}8\frac{3}{4}$	$9{\cdot}3\frac{3}{8}$	$10{\cdot}9\frac{7}{8}$	$12{\cdot}4\frac{1}{2}$	13·11	$15{\cdot}5\frac{1}{2}$

$32\frac{1}{2}°$ PITCH

RISE OF COMMON RAFTER $7\frac{5}{8}''$ PER FOOT OF RUN

BEVELS:

1.	COMMON RAFTER	–	SEAT	$32\frac{1}{2}$
2.	" "	–	RIDGE	$57\frac{1}{2}$
3.	HIP OR VALLEY	–	SEAT	$24\frac{1}{2}$
4.	" " "	–	RIDGE	$65\frac{3}{4}$
5.	JACK RAFTER	–	EDGE	$40\frac{1}{4}$
6.	PURLIN	–	EDGE	$49\frac{3}{4}$
7.	"	–	SIDE	$61\frac{1}{2}$

JACK RAFTERS 16 in. CENTRES DECREASE 19″, 18 in.—$21\frac{3}{8}''$, 24 in.—2′ $4\frac{1}{2}''$

RUN OF RAFTER *ins.*	$\frac{1}{2}$	1	2	3	4	5	6	7	8	9	10	11
LENGTH OF RAFTER	$\frac{5}{8}$	$1\frac{1}{8}$	$2\frac{3}{8}$	$3\frac{1}{2}$	$4\frac{3}{4}$	$5\frac{7}{8}$	$7\frac{1}{8}$	$8\frac{1}{4}$	$9\frac{1}{2}$	$10\frac{5}{8}$	$11\frac{7}{8}$	13
LENGTH OF HIP	$\frac{3}{4}$	$1\frac{1}{2}$	$3\frac{1}{8}$	$4\frac{5}{8}$	$6\frac{1}{8}$	$7\frac{3}{4}$	$9\frac{1}{4}$	$10\frac{7}{8}$	$12\frac{3}{8}$	14	$15\frac{1}{2}$	17

RUN OF RAFTER *ft.*	1	2	3	4	5	6	7	8	9	10
LENGTH OF RAFTER	$1{\cdot}2\frac{1}{4}$	$2{\cdot}4\frac{1}{2}$	$3{\cdot}6\frac{5}{8}$	$4{\cdot}8\frac{7}{8}$	$5{\cdot}11\frac{1}{8}$	$7{\cdot}1\frac{3}{8}$	$8{\cdot}3\frac{5}{8}$	$9{\cdot}5\frac{7}{8}$	10·8	$11{\cdot}10\frac{1}{4}$
LENGTH OF HIP	$1{\cdot}6\frac{5}{8}$	$3{\cdot}1\frac{1}{4}$	$4{\cdot}7\frac{7}{8}$	$6{\cdot}2\frac{3}{8}$	7·9	$9{\cdot}3\frac{5}{8}$	$10{\cdot}10\frac{1}{4}$	$12{\cdot}4\frac{7}{8}$	$13{\cdot}11\frac{1}{2}$	$15{\cdot}6\frac{1}{8}$

33° PITCH

RISE OF COMMON RAFTER $7\frac{13}{16}''$ PER FOOT OF RUN

BEVELS:

1.	COMMON RAFTER	–	SEAT	33
2.	" "	–	RIDGE	57
3.	HIP OR VALLEY	–	SEAT	$24\frac{1}{2}$
4.	" " "	–	RIDGE	$65\frac{1}{2}$
5.	JACK RAFTER	–	EDGE	40
6.	PURLIN	–	EDGE	50
7.	"	–	SIDE	$61\frac{1}{2}$

JACK RAFTERS 16 in. CENTRES DECREASE $19\frac{1}{8}''$, 18 in.—$21\frac{1}{2}''$, 24 in.—$2'\ 4\frac{5}{8}''$

RUN OF RAFTER *ins.*	$\frac{1}{2}$	1	2	3	4	5	6	7	8	9	10	11
LENGTH OF RAFTER	$\frac{5}{8}$	$1\frac{1}{4}$	$2\frac{3}{8}$	$3\frac{5}{8}$	$4\frac{3}{4}$	6	$7\frac{1}{8}$	$8\frac{3}{8}$	$9\frac{1}{2}$	$10\frac{3}{4}$	$11\frac{7}{8}$	$13\frac{1}{8}$
LENGTH OF HIP	$\frac{3}{4}$	$1\frac{1}{2}$	$3\frac{1}{8}$	$4\frac{5}{8}$	$6\frac{1}{4}$	$7\frac{3}{4}$	$9\frac{3}{8}$	$10\frac{7}{8}$	$12\frac{1}{2}$	14	$15\frac{1}{2}$	$17\frac{1}{8}$

RUN OF RAFTER *ft.*	1	2	3	4	5	6	7	8	9	10
LENGTH OF RAFTER	$1{\cdot}2\frac{1}{4}$	$2{\cdot}4\frac{5}{8}$	$3{\cdot}6\frac{7}{8}$	$4{\cdot}9\frac{1}{4}$	$5{\cdot}11\frac{1}{2}$	$7{\cdot}1\frac{7}{8}$	$8{\cdot}4\frac{1}{8}$	$9{\cdot}6\frac{1}{2}$	$10{\cdot}8\frac{3}{4}$	$11{\cdot}11\frac{1}{8}$
LENGTH OF HIP	$1{\cdot}6\frac{5}{8}$	$3{\cdot}1\frac{3}{8}$	4·8	$6{\cdot}2\frac{3}{8}$	$7{\cdot}9\frac{3}{8}$	9·4	$10{\cdot}10\frac{5}{8}$	$12{\cdot}5\frac{3}{8}$	14·0	$15{\cdot}6\frac{5}{8}$

Note: Third Pitch (Rise = $\frac{1}{3}$ Span) has a rafter seat bevel of 33° 40′. Lengths of rafters, etc., are based on this angle

THIRD PITCH

RISE OF COMMON RAFTER 8″ PER FOOT OF RUN

BEVELS:				
	1. COMMON RAFTER	–	SEAT	$33\frac{1}{2}$
	2. ″ ″	–	RIDGE	$56\frac{1}{2}$
	3. HIP OR VALLEY	–	SEAT	25
	4. ″ ″ ″	–	RIDGE	65
	5. JACK RAFTER	–	EDGE	40
	6. PURLIN	–	EDGE	50
	7. ″	–	SIDE	61

JACK RAFTERS 16 in. CENTRES DECREASE $19\frac{1}{4}$″, 18 in.—$21\frac{5}{8}$″, 24 in.—2′ $4\frac{7}{8}$″

RUN OF RAFTER *ins.*	$\frac{1}{2}$	1	2	3	4	5	6	7	8	9	10	11
LENGTH OF RAFTER	$\frac{5}{8}$	$1\frac{1}{4}$	$2\frac{3}{8}$	$3\frac{5}{8}$	$4\frac{3}{4}$	6	$7\frac{1}{4}$	$8\frac{3}{8}$	$9\frac{5}{8}$	$10\frac{7}{8}$	12	$13\frac{1}{4}$
LENGTH OF HIP	$\frac{3}{4}$	$1\frac{1}{2}$	$3\frac{1}{8}$	$4\frac{5}{8}$	$6\frac{1}{4}$	$7\frac{7}{8}$	$9\frac{3}{8}$	$10\frac{7}{8}$	$12\frac{1}{2}$	$14\frac{1}{8}$	$15\frac{5}{8}$	$17\frac{1}{4}$

RUN OF RAFTER *ft.*	1	2	3	4	5	6	7	8	9	10
LENGTH OF RAFTER	$1{\cdot}2\frac{3}{8}$	$2{\cdot}4\frac{7}{8}$	$3{\cdot}7\frac{1}{4}$	$4{\cdot}9\frac{5}{8}$	$6{\cdot}0\frac{1}{8}$	$7{\cdot}2\frac{1}{2}$	$8{\cdot}4\frac{7}{8}$	$9{\cdot}7\frac{3}{8}$	$10{\cdot}9\frac{3}{4}$	$12{\cdot}0\frac{1}{8}$
LENGTH OF HIP	$1{\cdot}6\frac{3}{4}$	$3{\cdot}1\frac{1}{2}$	$4{\cdot}8\frac{1}{4}$	6·3	$7{\cdot}9\frac{3}{4}$	$9{\cdot}4\frac{1}{2}$	$10{\cdot}11\frac{1}{4}$	12·6	$14{\cdot}0\frac{7}{8}$	$15{\cdot}7\frac{5}{8}$

35° PITCH

RISE OF COMMON RAFTER $8\frac{7}{16}''$ PER FOOT OF RUN

BEVELS:				
1.	COMMON RAFTER	–	SEAT	35
2.	" "	–	RIDGE	55
3.	HIP OR VALLEY	–	SEAT	$26\frac{1}{2}$
4.	" " "	–	RIDGE	$63\frac{1}{2}$
5.	JACK RAFTER	–	EDGE	$39\frac{1}{2}$
6.	PURLIN	–	EDGE	$50\frac{1}{2}$
7.	"	–	SIDE	60

JACK RAFTERS 16 in. CENTRES DECREASE $19\frac{1}{2}''$, 18 in.—22″, 24 in.—2′ $5\frac{1}{4}''$

RUN OF RAFTER *ins.*	$\frac{1}{2}$	1	2	3	4	5	6	7	8	9	10	11
LENGTH OF RAFTER	$\frac{5}{8}$	$1\frac{1}{4}$	$2\frac{1}{2}$	$3\frac{5}{8}$	$4\frac{7}{8}$	$6\frac{1}{8}$	$7\frac{3}{8}$	$8\frac{1}{2}$	$9\frac{3}{4}$	11	$12\frac{1}{4}$	$13\frac{3}{8}$
LENGTH OF HIP	$\frac{3}{4}$	$1\frac{5}{8}$	$3\frac{1}{8}$	$4\frac{3}{4}$	$6\frac{3}{8}$	$7\frac{7}{8}$	$9\frac{1}{2}$	11	$12\frac{5}{8}$	$14\frac{1}{4}$	$15\frac{3}{4}$	$17\frac{3}{8}$

RUN OF RAFTER *ft.*	1	2	3	4	5	6	7	8	9	10
LENGTH OF RAFTER	$1{\cdot}2\frac{5}{8}$	$2{\cdot}5\frac{1}{4}$	3·8	$4{\cdot}10\frac{5}{8}$	$6{\cdot}1\frac{1}{4}$	$7{\cdot}3\frac{7}{8}$	$8{\cdot}6\frac{1}{2}$	$9{\cdot}9\frac{1}{4}$	$10{\cdot}11\frac{7}{8}$	$12{\cdot}2\frac{1}{2}$
LENGTH OF HIP	$1{\cdot}6\frac{7}{8}$	$3{\cdot}1\frac{7}{8}$	$4{\cdot}8\frac{3}{4}$	$6{\cdot}3\frac{3}{4}$	$7{\cdot}10\frac{5}{8}$	$9{\cdot}5\frac{5}{8}$	$11{\cdot}0\frac{1}{2}$	$12{\cdot}7\frac{1}{2}$	$14{\cdot}2\frac{1}{2}$	$15{\cdot}9\frac{3}{8}$

36° PITCH

RISE OF COMMON RAFTER $8\frac{3}{4}''$ PER FOOT OF RUN

BEVELS: 1.	COMMON RAFTER	–	SEAT	36
2.	" "	–	RIDGE	54
3.	HIP OR VALLEY	–	SEAT	27
4.	" " "	–	RIDGE	63
5.	JACK RAFTER	–	EDGE	39
6.	PURLIN	–	EDGE	51
7.	"	–	SIDE	$59\frac{1}{2}$

JACK RAFTERS 16 in. CENTRES DECREASE $19\frac{3}{4}''$, 18 in.—$22\frac{1}{4}''$, 24 in.—2′ $5\frac{5}{8}''$

RUN OF RAFTER *ins.*	$\frac{1}{2}$	1	2	3	4	5	6	7	8	9	10	11
LENGTH OF RAFTER	$\frac{5}{8}$	$1\frac{1}{4}$	$2\frac{1}{2}$	$3\frac{3}{4}$	5	$6\frac{1}{8}$	$7\frac{3}{8}$	$8\frac{5}{8}$	$9\frac{7}{8}$	$11\frac{1}{8}$	$12\frac{3}{8}$	$13\frac{5}{8}$
LENGTH OF HIP	$\frac{3}{4}$	$1\frac{5}{8}$	$3\frac{1}{8}$	$4\frac{3}{4}$	$6\frac{3}{8}$	8	$9\frac{1}{2}$	$11\frac{1}{8}$	$12\frac{3}{4}$	$14\frac{1}{4}$	$15\frac{7}{8}$	$17\frac{1}{2}$

RUN OF RAFTER *ft.*	1	2	3	4	5	6	7	8	9	10
LENGTH OF RAFTER	$1{\cdot}2\frac{7}{8}$	$2{\cdot}5\frac{5}{8}$	$3{\cdot}8\frac{1}{2}$	$4{\cdot}11\frac{3}{8}$	$6{\cdot}2\frac{1}{8}$	7·5	$8{\cdot}7\frac{7}{8}$	$9{\cdot}10\frac{5}{8}$	$11{\cdot}1\frac{1}{2}$	$12{\cdot}4\frac{3}{8}$
LENGTH OF HIP	$1{\cdot}7\frac{1}{8}$	$3{\cdot}2\frac{1}{8}$	$4{\cdot}9\frac{1}{4}$	$6{\cdot}4\frac{1}{4}$	$7{\cdot}11\frac{3}{8}$	$9{\cdot}6\frac{1}{2}$	$11{\cdot}1\frac{1}{2}$	$12{\cdot}8\frac{5}{8}$	$14{\cdot}3\frac{5}{8}$	$15{\cdot}10\frac{3}{4}$

37° PITCH

RISE OF COMMON RAFTER $9\frac{1}{16}''$ PER FOOT OF RUN

BEVELS:

1.	COMMON RAFTER	–	SEAT	37
2.	" "	–	RIDGE	53
3.	HIP OR VALLEY	–	SEAT	28
4.	" " "	–	RIDGE	62
5.	JACK RAFTER	–	EDGE	$38\frac{1}{2}$
6.	PURLIN	–	EDGE	$51\frac{1}{2}$
7.	"	–	SIDE	59

JACK RAFTERS 16 in. CENTRES DECREASE 20″, 18 in.—$22\frac{1}{2}''$, 24 in.—2′ 6″

RUN OF RAFTER *ins.*	$\frac{1}{2}$	1	2	3	4	5	6	7	8	9	10	11
LENGTH OF RAFTER	$\frac{5}{8}$	$1\frac{1}{4}$	$2\frac{1}{2}$	$3\frac{3}{4}$	5	$6\frac{1}{4}$	$7\frac{1}{2}$	$8\frac{3}{4}$	10	$11\frac{1}{4}$	$12\frac{1}{2}$	$13\frac{3}{4}$
LENGTH OF HIP	$\frac{3}{4}$	$1\frac{5}{8}$	$3\frac{1}{4}$	$4\frac{3}{4}$	$6\frac{3}{8}$	8	$9\frac{5}{8}$	$11\frac{1}{4}$	$12\frac{7}{8}$	$14\frac{3}{8}$	16	$17\frac{5}{8}$

RUN OF RAFTER *ft.*	1	2	3	4	5	6	7	8	9	10
LENGTH OF RAFTER	1·3	2·6	$3{\cdot}9\frac{1}{8}$	$5{\cdot}0\frac{1}{8}$	$6{\cdot}3\frac{1}{8}$	$7{\cdot}6\frac{1}{8}$	$8{\cdot}9\frac{1}{8}$	$10{\cdot}0\frac{1}{4}$	$11{\cdot}3\frac{1}{4}$	$12{\cdot}6\frac{1}{4}$
LENGTH OF HIP	$1{\cdot}7\frac{1}{4}$	$3{\cdot}2\frac{1}{2}$	$4{\cdot}9\frac{5}{8}$	$6{\cdot}4\frac{7}{8}$	$8{\cdot}0\frac{1}{8}$	$9{\cdot}7\frac{3}{8}$	$11{\cdot}2\frac{5}{8}$	$12{\cdot}9\frac{7}{8}$	14·5	$16{\cdot}0\frac{1}{4}$

$37\frac{1}{2}$° PITCH

RISE OF COMMON RAFTER $9\frac{3}{16}$″ PER FOOT OF RUN

BEVELS:				
1.	COMMON RAFTER	–	SEAT	$37\frac{1}{2}$
2.	″ ″	–	RIDGE	$52\frac{1}{2}$
3.	HIP OR VALLEY	–	SEAT	$28\frac{1}{2}$
4.	″ ″ ″	–	RIDGE	$61\frac{1}{2}$
5.	JACK RAFTER	–	EDGE	$38\frac{1}{4}$
6.	PURLIN	–	EDGE	$51\frac{3}{4}$
7.	″	–	SIDE	$58\frac{1}{2}$

JACK RAFTERS 16 in. CENTRES DECREASE $20\frac{1}{8}$″, 18 in.—$22\frac{3}{4}$″, 24 in.—2′ $6\frac{1}{4}$″

RUN OF RAFTER *ins.*	$\frac{1}{2}$	1	2	3	4	5	6	7	8	9	10	11
LENGTH OF RAFTER	$\frac{5}{8}$	$1\frac{1}{4}$	$2\frac{1}{2}$	$3\frac{3}{4}$	5	$6\frac{1}{4}$	$7\frac{1}{2}$	$8\frac{3}{4}$	10	$11\frac{3}{8}$	$12\frac{5}{8}$	14
LENGTH OF HIP	$\frac{3}{4}$	$1\frac{5}{8}$	$3\frac{1}{4}$	$4\frac{3}{4}$	$6\frac{3}{8}$	8	$9\frac{5}{8}$	$11\frac{1}{4}$	$12\frac{7}{8}$	$14\frac{1}{2}$	$16\frac{1}{8}$	$17\frac{3}{4}$

RUN OF RAFTER *ft.*	1	2	3	4	5	6	7	8	9	10
LENGTH OF RAFTER	$1{\cdot}3\frac{1}{8}$	$2{\cdot}6\frac{1}{4}$	$3{\cdot}9\frac{3}{8}$	$5{\cdot}0\frac{1}{2}$	$6{\cdot}3\frac{5}{8}$	$7{\cdot}6\frac{3}{4}$	$8{\cdot}9\frac{7}{8}$	10·1	$11{\cdot}4\frac{1}{8}$	$12{\cdot}7\frac{1}{4}$
LENGTH OF HIP	$1{\cdot}7\frac{3}{8}$	$3{\cdot}2\frac{5}{8}$	4·10	$6{\cdot}5\frac{1}{4}$	$8{\cdot}0\frac{1}{2}$	$9{\cdot}7\frac{7}{8}$	$11{\cdot}3\frac{1}{4}$	$12{\cdot}10\frac{1}{2}$	$14{\cdot}5\frac{3}{4}$	$16{\cdot}1\frac{1}{8}$

38° PITCH

RISE OF COMMON RAFTER $9\frac{3}{8}''$ PER FOOT OF RUN

BEVELS:				
1.	COMMON RAFTER	–	SEAT	38
2.	" "	–	RIDGE	52
3.	HIP OR VALLEY	–	SEAT	29
4.	" " "	–	RIDGE	61
5.	JACK RAFTER	–	EDGE	38
6.	PURLIN	–	EDGE	51
7.	"	–	SIDE	$58\frac{1}{2}$

JACK RAFTERS 16 in. CENTRES DECREASE $20\frac{1}{4}''$, 18 in.—$22\frac{7}{8}''$, 24 in.—2′ $6\frac{1}{2}''$

RUN OF RAFTER *ins.*	$\frac{1}{2}$	1	2	3	4	5	6	7	8	9	10	11
LENGTH OF RAFTER	$\frac{5}{8}$	$1\frac{1}{4}$	$2\frac{1}{2}$	$3\frac{3}{4}$	$5\frac{1}{8}$	$6\frac{3}{8}$	$7\frac{5}{8}$	$8\frac{7}{8}$	$10\frac{1}{8}$	$11\frac{3}{8}$	$12\frac{3}{4}$	14
LENGTH OF HIP	$\frac{3}{4}$	$1\frac{5}{8}$	$3\frac{1}{4}$	$4\frac{7}{8}$	$6\frac{1}{2}$	$8\frac{1}{8}$	$9\frac{3}{4}$	$11\frac{1}{4}$	$12\frac{7}{8}$	$14\frac{1}{2}$	$16\frac{1}{8}$	$17\frac{3}{4}$

RUN OF RAFTER *ft.*	1	2	3	4	5	6	7	8	9	10
LENGTH OF RAFTER	$1{\cdot}3\frac{1}{4}$	$2{\cdot}6\frac{1}{2}$	$3{\cdot}9\frac{5}{8}$	$5{\cdot}0\frac{7}{8}$	$6{\cdot}4\frac{1}{8}$	$7{\cdot}7\frac{3}{8}$	$8{\cdot}10\frac{5}{8}$	$10{\cdot}1\frac{7}{8}$	11·5	$12{\cdot}8\frac{1}{4}$
LENGTH OF HIP	$1{\cdot}7\frac{3}{8}$	$3{\cdot}2\frac{3}{4}$	$4{\cdot}10\frac{1}{8}$	$6{\cdot}5\frac{1}{2}$	$8{\cdot}0\frac{7}{8}$	$9{\cdot}8\frac{1}{4}$	$11{\cdot}3\frac{5}{8}$	12·11	$14{\cdot}6\frac{3}{8}$	$16{\cdot}1\frac{3}{4}$

39° PITCH

RISE OF COMMON RAFTER $9\frac{3}{4}''$ PER FOOT OF RUN

BEVELS:				
1.	COMMON RAFTER	–	SEAT	39
2.	" "	–	RIDGE	51
3.	HIP OR VALLEY	–	SEAT	30
4.	" " "	–	RIDGE	60
5.	JACK RAFTER	–	EDGE	38
6.	PURLIN	–	EDGE	52
7.	"	–	SIDE	58

JACK RAFTERS 16 in. CENTRES DECREASE $20\frac{5}{8}''$, 18 in.—$23\frac{1}{8}''$, 24 in.—$2'\ 6\frac{7}{8}''$

RUN OF RAFTER *ins.*	$\frac{1}{2}$	1	2	3	4	5	6	7	8	9	10	11
LENGTH OF RAFTER	$\frac{5}{8}$	$1\frac{1}{4}$	$2\frac{5}{8}$	$3\frac{7}{8}$	$5\frac{1}{8}$	$6\frac{3}{8}$	$7\frac{3}{4}$	9	$10\frac{1}{4}$	$11\frac{5}{8}$	$12\frac{7}{8}$	$14\frac{1}{8}$
LENGTH OF HIP	$\frac{3}{4}$	$1\frac{5}{8}$	$3\frac{1}{4}$	$4\frac{7}{8}$	$6\frac{1}{2}$	$8\frac{1}{8}$	$9\frac{3}{4}$	$11\frac{3}{8}$	13	$14\frac{5}{8}$	$16\frac{1}{4}$	$17\frac{7}{8}$

RUN OF RAFTER *ft.*	1	2	3	4	5	6	7	8	9	10
LENGTH OF RAFTER	$1{\cdot}3\frac{1}{2}$	$2{\cdot}6\frac{7}{8}$	$3{\cdot}10\frac{3}{8}$	$5{\cdot}1\frac{3}{4}$	$6{\cdot}5\frac{1}{4}$	$7{\cdot}8\frac{5}{8}$	$9{\cdot}0\frac{1}{8}$	$10{\cdot}3\frac{1}{2}$	11·7	$12{\cdot}10\frac{3}{8}$
LENGTH OF HIP	$1{\cdot}7\frac{1}{2}$	$3{\cdot}3\frac{1}{8}$	$4{\cdot}10\frac{5}{8}$	$6{\cdot}6\frac{1}{8}$	$8{\cdot}1\frac{3}{4}$	$9{\cdot}9\frac{1}{4}$	$11{\cdot}4\frac{7}{8}$	$13{\cdot}0\frac{3}{8}$	$14{\cdot}7\frac{7}{8}$	$16{\cdot}3\frac{1}{2}$

40° PITCH

RISE OF COMMON RAFTER 10$\frac{1}{16}$″ PER FOOT OF RUN

BEVELS:				
1.	COMMON RAFTER	–	SEAT	40
2.	″ ″	–	RIDGE	50
3.	HIP OR VALLEY	–	SEAT	30$\frac{1}{2}$
4.	″ ″ ″	–	RIDGE	59$\frac{1}{2}$
5.	JACK RAFTER	–	EDGE	37$\frac{1}{2}$
6.	PURLIN	–	EDGE	52$\frac{1}{2}$
7.	″	–	SIDE	57$\frac{1}{2}$

JACK RAFTERS 16 in. CENTRES DECREASE 20$\frac{7}{8}$″, 18 in.—23$\frac{1}{2}$″, 24 in.—2′ 7$\frac{3}{8}$″

RUN OF RAFTER *ins.*	$\frac{1}{2}$	1	2	3	4	5	6	7	8	9	10	11
LENGTH OF RAFTER	$\frac{5}{8}$	1$\frac{1}{4}$	2$\frac{5}{8}$	3$\frac{7}{8}$	5$\frac{1}{4}$	6$\frac{1}{2}$	7$\frac{7}{8}$	9$\frac{1}{8}$	10$\frac{1}{2}$	11$\frac{3}{4}$	13	14$\frac{3}{8}$
LENGTH OF HIP	$\frac{7}{8}$	1$\frac{5}{8}$	3$\frac{1}{4}$	4$\frac{7}{8}$	6$\frac{5}{8}$	8$\frac{1}{4}$	9$\frac{7}{8}$	11$\frac{1}{2}$	13$\frac{1}{8}$	14$\frac{3}{4}$	16$\frac{1}{2}$	18$\frac{1}{8}$

RUN OF RAFTER *ft.*	1	2	3	4	5	6	7	8	9	10
LENGTH OF RAFTER	1·3$\frac{5}{8}$	2·7$\frac{3}{8}$	3·11	5·2$\frac{5}{8}$	6·6$\frac{3}{8}$	7·10	9·1$\frac{5}{8}$	10·5$\frac{3}{8}$	11·9	13·0$\frac{5}{8}$
LENGTH OF HIP	1·7$\frac{3}{4}$	3·3$\frac{1}{2}$	4·11$\frac{1}{8}$	6·6$\frac{7}{8}$	8·2$\frac{5}{8}$	9·10$\frac{3}{8}$	11·6$\frac{1}{8}$	13·1$\frac{7}{8}$	14·9$\frac{1}{2}$	16·5$\frac{1}{4}$

41° PITCH

RISE OF COMMON RAFTER $10\frac{7}{16}''$ PER FOOT OF RUN

BEVELS:					
1.	COMMON RAFTER	–	SEAT	41
2.	" "	–	RIDGE	49
3.	HIP OR VALLEY	–	SEAT	$31\frac{1}{2}$
4.	" " "	–	RIDGE	$58\frac{1}{2}$
5.	JACK RAFTER	–	EDGE	37
6.	PURLIN	–	EDGE	53
7.	"	–	SIDE	$56\frac{1}{2}$

JACK RAFTERS 16 in. CENTRES DECREASE $21\frac{1}{4}''$, 18 in.—$23\frac{7}{8}''$, 24 in.—2′ $7\frac{3}{4}''$

RUN OF RAFTER *ins.*	$\frac{1}{2}$	1	2	3	4	5	6	7	8	9	10	11
LENGTH OF RAFTER	$\frac{5}{8}$	$1\frac{3}{8}$	$2\frac{5}{8}$	4	$5\frac{1}{4}$	$6\frac{5}{8}$	8	$9\frac{1}{4}$	$10\frac{5}{8}$	$11\frac{7}{8}$	$13\frac{1}{4}$	$14\frac{5}{8}$
LENGTH OF HIP	$\frac{7}{8}$	$1\frac{5}{8}$	$3\frac{3}{8}$	5	$6\frac{5}{8}$	$8\frac{1}{4}$	10	$11\frac{5}{8}$	$13\frac{1}{4}$	15	$16\frac{5}{8}$	$18\frac{1}{4}$

RUN OF RAFTER *ft.*	1	2	3	4	5	6	7	8	9	10
LENGTH OF RAFTER	$1{\cdot}3\frac{7}{8}$	$2{\cdot}7\frac{3}{4}$	$3{\cdot}11\frac{3}{4}$	$5{\cdot}3\frac{5}{8}$	$6{\cdot}7\frac{1}{2}$	$7{\cdot}11\frac{3}{8}$	$9{\cdot}3\frac{1}{4}$	$10{\cdot}7\frac{1}{4}$	$11{\cdot}11\frac{1}{8}$	13·3
LENGTH OF HIP	$1{\cdot}7\frac{7}{8}$	$3{\cdot}3\frac{7}{8}$	$4{\cdot}11\frac{3}{4}$	$6{\cdot}7\frac{5}{8}$	$8{\cdot}3\frac{5}{8}$	$9{\cdot}11\frac{1}{2}$	$11{\cdot}7\frac{1}{2}$	$13{\cdot}3\frac{3}{8}$	$14{\cdot}11\frac{1}{4}$	$16{\cdot}7\frac{1}{4}$

42° PITCH

RISE OF COMMON RAFTER $10\frac{13}{16}''$ PER FOOT OF RUN

BEVELS:				
1.	COMMON RAFTER	–	SEAT	42
2.	″ ″	–	RIDGE	48
3.	HIP OR VALLEY	–	SEAT	$32\frac{1}{2}$
4.	″ ″ ″	–	RIDGE	$57\frac{1}{2}$
5.	JACK RAFTER	–	EDGE	$36\frac{1}{2}$
6.	PURLIN	–	EDGE	$53\frac{1}{2}$
7.	″	–	SIDE	56

JACK RAFTERS 16 in. CENTRES DECREASE $21\frac{1}{2}''$, 18 in.—$2'\ 0\frac{1}{4}''$, 24 in.—$2'\ 8\frac{1}{4}''$

RUN OF RAFTER *ins.*	$\frac{1}{2}$	1	2	3	4	5	6	7	8	9	10	11
LENGTH OF RAFTER	$\frac{5}{8}$	$1\frac{3}{8}$	$2\frac{3}{4}$	4	$5\frac{3}{8}$	$6\frac{3}{4}$	$8\frac{1}{8}$	$9\frac{3}{8}$	$10\frac{3}{4}$	$12\frac{1}{8}$	$13\frac{1}{2}$	$14\frac{3}{4}$
LENGTH OF HIP	$\frac{7}{8}$	$1\frac{5}{8}$	$3\frac{3}{8}$	5	$6\frac{3}{4}$	$8\frac{3}{8}$	10	$11\frac{3}{4}$	$13\frac{3}{8}$	$15\frac{1}{8}$	$16\frac{3}{4}$	$18\frac{3}{8}$

RUN OF RAFTER *ft.*	1	2	3	4	5	6	7	8	9	10
LENGTH OF RAFTER	$1{\cdot}4\frac{1}{8}$	$2{\cdot}8\frac{1}{4}$	$4{\cdot}0\frac{1}{2}$	$5{\cdot}4\frac{5}{8}$	$6{\cdot}8\frac{3}{4}$	$8{\cdot}0\frac{7}{8}$	9·5	$10{\cdot}9\frac{1}{8}$	$12{\cdot}1\frac{3}{8}$	$13{\cdot}5\frac{1}{2}$
LENGTH OF HIP	$1{\cdot}8\frac{1}{8}$	$3{\cdot}4\frac{1}{4}$	$5{\cdot}0\frac{3}{8}$	$6{\cdot}8\frac{1}{2}$	$8{\cdot}4\frac{5}{8}$	$10{\cdot}0\frac{3}{4}$	$11{\cdot}8\frac{7}{8}$	13·5	15·1	$16{\cdot}9\frac{1}{8}$

$42\frac{1}{2}°$ PITCH

RISE OF COMMON RAFTER 11″ PER FOOT OF RUN

BEVELS:

1.	COMMON RAFTER	–	SEAT	$42\frac{1}{2}$
2.	″ ″	–	RIDGE	$47\frac{1}{2}$
3.	HIP OR VALLEY	–	SEAT	33
4.	″ ″ ″	–	RIDGE	57
5.	JACK RAFTER	–	EDGE	$36\frac{1}{4}$
6.	PURLIN	–	EDGE	$53\frac{3}{4}$
7.	″	–	SIDE	$55\frac{3}{4}$

JACK RAFTERS 16 in. CENTRES DECREASE $21\frac{3}{4}$″, 18 in.—2′ $0\frac{3}{8}$″, 24 in.—2′ $8\frac{5}{8}$″

RUN OF RAFTER *ins.*	$\frac{1}{2}$	1	2	3	4	5	6	7	8	9	10	11
LENGTH OF RAFTER	$\frac{5}{8}$	$1\frac{3}{8}$	$2\frac{3}{4}$	4	$5\frac{3}{8}$	$6\frac{3}{4}$	$8\frac{1}{8}$	$9\frac{3}{8}$	$10\frac{3}{4}$	$12\frac{1}{4}$	$13\frac{5}{8}$	$14\frac{7}{8}$
LENGTH OF HIP	$\frac{7}{8}$	$1\frac{5}{8}$	$3\frac{3}{8}$	5	$6\frac{3}{4}$	$8\frac{3}{8}$	10	$11\frac{3}{4}$	$13\frac{3}{8}$	$15\frac{1}{8}$	$16\frac{7}{8}$	$18\frac{1}{2}$

RUN OF RAFTER *ft.*	1	2	3	4	5	6	7	8	9	10
LENGTH OF RAFTER	$1{\cdot}4\frac{1}{4}$	$2{\cdot}8\frac{1}{2}$	$4{\cdot}0\frac{7}{8}$	$5{\cdot}5\frac{1}{8}$	$6{\cdot}9\frac{3}{8}$	$8{\cdot}1\frac{5}{8}$	9·6	$10{\cdot}10\frac{1}{4}$	$12{\cdot}2\frac{1}{2}$	$13{\cdot}6\frac{3}{4}$
LENGTH OF HIP	$1{\cdot}8\frac{1}{4}$	$3{\cdot}4\frac{1}{2}$	$5{\cdot}0\frac{5}{8}$	6·9	$8{\cdot}5\frac{1}{8}$	$10{\cdot}1\frac{1}{8}$	$11{\cdot}9\frac{5}{8}$	$13{\cdot}5\frac{7}{8}$	15·2	$16{\cdot}10\frac{3}{8}$

43° PITCH

RISE OF COMMON RAFTER 11$\frac{3}{16}$″ PER FOOT OF RUN

BEVELS:
1. COMMON RAFTER – SEAT 43
2. ″ ″ – RIDGE 47
3. HIP OR VALLEY – SEAT 33$\frac{1}{2}$
4. ″ ″ ″ – RIDGE 56$\frac{1}{2}$
5. JACK RAFTER – EDGE 36
6. PURLIN – EDGE 54
7. ″ – SIDE 55$\frac{1}{2}$

JACK RAFTERS 16 in. CENTRES DECREASE 21$\frac{7}{8}$″, 18 in.—2′ 0$\frac{5}{8}$″, 24 in.—2′ 8$\frac{7}{8}$″

RUN OF RAFTER *ins.*	$\frac{1}{2}$	1	2	3	4	5	6	7	8	9	10	11
LENGTH OF RAFTER	$\frac{5}{8}$	1$\frac{3}{8}$	2$\frac{3}{4}$	4$\frac{1}{8}$	5$\frac{1}{2}$	6$\frac{7}{8}$	8$\frac{1}{4}$	9$\frac{5}{8}$	10$\frac{7}{8}$	12$\frac{1}{4}$	13$\frac{5}{8}$	15
LENGTH OF HIP	$\frac{7}{8}$	1$\frac{3}{4}$	3$\frac{3}{8}$	5$\frac{1}{8}$	6$\frac{3}{4}$	8$\frac{1}{2}$	10$\frac{1}{8}$	11$\frac{7}{8}$	13$\frac{1}{2}$	15$\frac{1}{4}$	17	18$\frac{5}{8}$

RUN OF RAFTER *ft.*	1	2	3	4	5	6	7	8	9	10
LENGTH OF RAFTER	1·4$\frac{3}{8}$	2·8$\frac{7}{8}$	4·1$\frac{1}{4}$	5·5$\frac{5}{8}$	6·10	8·2$\frac{1}{2}$	9·6$\frac{7}{8}$	10·11$\frac{1}{4}$	12·3$\frac{5}{8}$	13·8$\frac{1}{8}$
LENGTH OF HIP	1·8$\frac{3}{8}$	3·4$\frac{5}{8}$	5·1	6·9$\frac{1}{4}$	8·5$\frac{5}{8}$	10·2	11·10$\frac{1}{4}$	13·6$\frac{5}{8}$	15·3	16·11$\frac{1}{4}$

44° PITCH

RISE OF COMMON RAFTER 11$\frac{9}{16}$″ PER FOOT OF RUN

BEVELS:

1.	COMMON RAFTER	–	SEAT	44
2.	″ ″	–	RIDGE	46
3.	HIP OR VALLEY	–	SEAT	34$\frac{1}{2}$
4.	″ ″ ″	–	RIDGE	55$\frac{1}{2}$
5.	JACK RAFTER	–	EDGE	35$\frac{1}{2}$
6.	PURLIN	–	EDGE	54$\frac{1}{2}$
7.	″	–	SIDE	55

JACK RAFTERS 16 in. CENTRES DECREASE 22$\frac{1}{4}$″, 18 in.—2′ 1″, 24 in.—2′ 9$\frac{3}{8}$″

RUN OF RAFTER *ins.*	$\frac{1}{2}$	1	2	3	4	5	6	7	8	9	10	11
LENGTH OF RAFTER	$\frac{3}{4}$	1$\frac{3}{8}$	2$\frac{3}{4}$	4$\frac{1}{8}$	5$\frac{1}{2}$	7	8$\frac{3}{8}$	9$\frac{3}{4}$	11$\frac{1}{8}$	12$\frac{1}{2}$	13$\frac{7}{8}$	15$\frac{1}{4}$
LENGTH OF HIP	$\frac{7}{8}$	1$\frac{3}{4}$	3$\frac{3}{8}$	5$\frac{1}{8}$	6$\frac{7}{8}$	8$\frac{1}{2}$	10$\frac{1}{4}$	12	13$\frac{1}{4}$	15$\frac{3}{8}$	17$\frac{1}{8}$	18$\frac{7}{8}$

RUN OF RAFTER *ft.*	1	2	3	4	5	6	7	8	9	10
LENGTH OF RAFTER	1·4$\frac{5}{8}$	2·9$\frac{3}{8}$	4·2$\frac{1}{4}$	5·6$\frac{3}{4}$	6·11$\frac{3}{8}$	8·4$\frac{1}{8}$	9·8$\frac{3}{4}$	11·1$\frac{1}{2}$	12·6$\frac{1}{8}$	13·10$\frac{7}{8}$
LENGTH OF HIP	1·8$\frac{1}{2}$	3·5$\frac{1}{8}$	5·1$\frac{5}{8}$	6·10$\frac{1}{8}$	8·6$\frac{3}{4}$	10·3$\frac{1}{4}$	11·11$\frac{3}{4}$	13·8$\frac{3}{8}$	15·4$\frac{7}{8}$	17·1$\frac{3}{8}$

45° or TRUE PITCH

RISE OF COMMON RAFTER 12″ PER FOOT OF RUN

BEVELS:				
1.	COMMON RAFTER	–	SEAT	45
2.	″ ″	–	RIDGE	45
3.	HIP OR VALLEY	–	SEAT	$35\frac{1}{2}$
4.	″ ″ ″	–	RIDGE	$54\frac{1}{2}$
5.	JACK RAFTER	–	EDGE	$35\frac{1}{2}$
6.	PURLIN	–	EDGE	$54\frac{1}{2}$
7.	″	–	SIDE	$54\frac{1}{2}$

JACK RAFTERS 16 in. CENTRES DECREASE $22\frac{5}{8}''$, 18 in.—2′ $1\frac{1}{2}''$, 24 in.—2′ 10″

RUN OF RAFTER *ins.*	$\frac{1}{2}$	1	2	3	4	5	6	7	8	9	10	11
LENGTH OF RAFTER	$\frac{3}{4}$	$1\frac{3}{8}$	$2\frac{7}{8}$	$4\frac{1}{4}$	$5\frac{5}{8}$	$7\frac{1}{8}$	$8\frac{1}{2}$	$9\frac{7}{8}$	$11\frac{1}{4}$	$12\frac{3}{4}$	$14\frac{1}{8}$	$15\frac{1}{2}$
LENGTH OF HIP	$\frac{7}{8}$	$1\frac{3}{4}$	$3\frac{1}{2}$	$5\frac{1}{4}$	$6\frac{7}{8}$	$8\frac{5}{8}$	$10\frac{3}{8}$	$12\frac{1}{8}$	$13\frac{7}{8}$	$15\frac{5}{8}$	$17\frac{3}{8}$	19

RUN OF RAFTER *ft.*	1	2	3	4	5	6	7	8	9	10
LENGTH OF RAFTER	1·5	2·10	4·$2\frac{7}{8}$	5·$7\frac{5}{8}$	7·$0\frac{7}{8}$	8·$5\frac{7}{8}$	9·$10\frac{3}{4}$	11·$3\frac{3}{4}$	12·$8\frac{3}{4}$	14·$1\frac{3}{4}$
LENGTH OF HIP	1·$8\frac{3}{4}$	3·$5\frac{1}{2}$	5·$2\frac{3}{8}$	6·$11\frac{1}{8}$	8·$7\frac{7}{8}$	10·$4\frac{3}{4}$	12·$1\frac{1}{2}$	13·$10\frac{1}{4}$	15·7	17·$3\frac{7}{8}$

46° PITCH

RISE OF COMMON RAFTER 1′ 0$\frac{7}{16}$″ PER FOOT OF RUN

BEVELS:			
1.	COMMON RAFTER	– SEAT	46
2.	″ ″	– RIDGE	44
3.	HIP OR VALLEY	– SEAT	36
4.	″ ″ ″	– RIDGE	54
5.	JACK RAFTER	– EDGE	35
6.	PURLIN	– EDGE	55
7.	″	– SIDE	54$\frac{1}{2}$

JACK RAFTERS 16 in. CENTRES DECREASE 23″, 18 in.—2′ 1$\frac{7}{8}$″, 24 in.—2′ 10$\frac{1}{2}$″

RUN OF RAFTER *ins.*	$\frac{1}{2}$	1	2	3	4	5	6	7	8	9	10	11
LENGTH OF RAFTER	$\frac{3}{4}$	1$\frac{1}{2}$	2$\frac{7}{8}$	4$\frac{3}{8}$	5$\frac{3}{4}$	7$\frac{1}{4}$	8$\frac{5}{8}$	10$\frac{1}{8}$	11$\frac{1}{2}$	13	14$\frac{3}{8}$	15$\frac{7}{8}$
LENGTH OF HIP	$\frac{7}{8}$	1$\frac{3}{4}$	3$\frac{1}{2}$	5$\frac{1}{4}$	7	8$\frac{3}{4}$	10$\frac{1}{2}$	12$\frac{1}{4}$	14	15$\frac{3}{4}$	17$\frac{1}{2}$	19$\frac{1}{4}$

RUN OF RAFTER *ft.*	1	2	3	4	5	6	7	8	9	10
LENGTH OF RAFTER	1·5$\frac{1}{4}$	2·10$\frac{1}{2}$	4·3$\frac{7}{8}$	5·9$\frac{1}{8}$	7·2$\frac{3}{8}$	8·7$\frac{5}{8}$	10·0$\frac{7}{8}$	11·6$\frac{1}{4}$	12·11$\frac{1}{2}$	14·4$\frac{3}{4}$
LENGTH OF HIP	1·9	3·6$\frac{1}{8}$	5·3$\frac{1}{8}$	7·0$\frac{1}{8}$	8·9$\frac{1}{8}$	10·6$\frac{1}{4}$	12·3$\frac{1}{4}$	14·0$\frac{1}{4}$	15·9$\frac{1}{4}$	17·6$\frac{3}{8}$

47° PITCH

RISE OF COMMON RAFTER 1′ 0$\frac{7}{8}$″ PER FOOT OF RUN

BEVELS:				
	1. COMMON RAFTER	–	SEAT	47
	2. ″ ″	–	RIDGE	43
	3. HIP OR VALLEY	–	SEAT	37
	4. ″ ″ ″	–	RIDGE	53
	5. JACK RAFTER	–	EDGE	34$\frac{1}{2}$
	6. PURLIN	–	EDGE	55$\frac{1}{2}$
	7. ″	–	SIDE	54

JACK RAFTERS 16 in. CENTRES DECREASE 23$\frac{1}{2}$″, 18 in.—2′ 2$\frac{3}{8}$″, 24 in.—2′ 11$\frac{1}{4}$″

RUN OF RAFTER *ins.*	$\frac{1}{2}$	1	2	3	4	5	6	7	8	9	10	11
LENGTH OF RAFTER	$\frac{3}{4}$	1$\frac{1}{2}$	2$\frac{7}{8}$	4$\frac{3}{8}$	5$\frac{7}{8}$	7$\frac{3}{8}$	8$\frac{3}{4}$	10$\frac{1}{4}$	11$\frac{3}{4}$	13$\frac{1}{4}$	14$\frac{5}{8}$	16$\frac{1}{8}$
LENGTH OF HIP	$\frac{7}{8}$	1$\frac{3}{4}$	3$\frac{1}{2}$	5$\frac{3}{8}$	7$\frac{1}{8}$	8$\frac{7}{8}$	10$\frac{5}{8}$	12$\frac{3}{8}$	14$\frac{1}{4}$	16	17$\frac{3}{4}$	19$\frac{1}{2}$

RUN OF RAFTER *ft.*	1	2	3	4	5	6	7	8	9	10
LENGTH OF RAFTER	1·5$\frac{5}{8}$	2·11$\frac{1}{4}$	4·4$\frac{3}{4}$	5·10$\frac{3}{8}$	7·4	8·9$\frac{5}{8}$	10·3$\frac{1}{8}$	11·8$\frac{3}{4}$	13·2$\frac{3}{8}$	14·8
LENGTH OF HIP	1·9$\frac{1}{4}$	3·6$\frac{5}{8}$	5·3$\frac{7}{8}$	7·1$\frac{1}{8}$	8·10$\frac{1}{2}$	10·7$\frac{3}{4}$	12·5$\frac{1}{8}$	14·2$\frac{5}{8}$	15·11$\frac{5}{6}$	17·9

48° PITCH

RISE OF COMMON RAFTER 1′ 1$\frac{5}{16}$″ PER FOOT OF RUN

BEVELS:				
1.	COMMON RAFTER	–	SEAT	48
2.	″ ″	–	RIDGE	42
3.	HIP OR VALLEY	–	SEAT	38
4.	″ ″ ″	–	RIDGE	52
5.	JACK RAFTER	–	EDGE	34
6.	PURLIN	–	EDGE	56
7.	″	–	SIDE	53$\frac{1}{2}$

JACK RAFTERS 16 in. CENTRES DECREASE 23$\frac{7}{8}$″, 18 in.—2′ 2$\frac{7}{8}$″, 24 in.—2′ 11$\frac{7}{8}$″

RUN OF RAFTER *ins.*	$\frac{1}{2}$	1	2	3	4	5	6	7	8	9	10	11
LENGTH OF RAFTER	$\frac{3}{4}$	1$\frac{1}{2}$	3	4$\frac{1}{2}$	6	7$\frac{1}{2}$	9	10$\frac{1}{2}$	12	13$\frac{1}{2}$	15	16$\frac{1}{2}$
LENGTH OF HIP	$\frac{7}{8}$	1$\frac{3}{4}$	3$\frac{5}{8}$	5$\frac{3}{8}$	7$\frac{1}{4}$	9	10$\frac{3}{4}$	12$\frac{5}{8}$	14$\frac{3}{8}$	16$\frac{1}{8}$	18	19$\frac{3}{4}$

RUN OF RAFTER *ft.*	1	2	3	4	5	6	7	8	9	10
LENGTH OF RAFTER	1·5$\frac{7}{8}$	2·11$\frac{7}{8}$	4·5$\frac{3}{4}$	5·11$\frac{3}{4}$	7·5$\frac{5}{8}$	8·11$\frac{5}{8}$	10·5$\frac{1}{2}$	11·11$\frac{1}{2}$	13·5$\frac{3}{8}$	14·11$\frac{3}{8}$
LENGTH OF HIP	1·9$\frac{5}{8}$	3·7$\frac{1}{8}$	5·4$\frac{3}{4}$	7·2$\frac{1}{4}$	8·11$\frac{7}{8}$	10·9$\frac{1}{2}$	12·7	14·4$\frac{5}{8}$	16·2$\frac{1}{8}$	17·11$\frac{3}{4}$

49° PITCH

RISE OF COMMON RAFTER 1′ 1$\frac{13}{16}$″ PER FOOT OF RUN

BEVELS:				
1.	COMMON RAFTER	–	SEAT	49
2.	″ ″	–	RIDGE	41
3.	HIP OR VALLEY	–	SEAT	39
4.	″ ″ ″	–	RIDGE	51
5.	JACK RAFTER	–	EDGE	33$\frac{1}{2}$
6.	PURLIN	–	EDGE	56$\frac{1}{2}$
7.	″	–	SIDE	53

JACK RAFTERS 16 in. CENTRES DECREASE 2′ 0$\frac{3}{8}$″, 18 in.—2′ 3$\frac{3}{8}$″, 24 in.—3′ 0$\frac{5}{8}$″

RUN OF RAFTER *ins.*	$\frac{1}{2}$	1	2	3	4	5	6	7	8	9	10	11
LENGTH OF RAFTER	$\frac{3}{4}$	1$\frac{1}{2}$	3	4$\frac{5}{8}$	6$\frac{1}{8}$	7$\frac{5}{8}$	9$\frac{1}{8}$	10$\frac{5}{8}$	12$\frac{1}{4}$	13$\frac{3}{4}$	15$\frac{1}{4}$	16$\frac{3}{4}$
LENGTH OF HIP	$\frac{7}{8}$	1$\frac{7}{8}$	3$\frac{5}{8}$	5$\frac{1}{2}$	7$\frac{1}{4}$	9$\frac{1}{8}$	10$\frac{7}{8}$	12$\frac{3}{4}$	14$\frac{5}{8}$	16$\frac{3}{8}$	18$\frac{1}{4}$	20

RUN OF RAFTER *ft.*	1	2	3	4	5	6	7	8	9	10
LENGTH OF RAFTER	1·6$\frac{1}{4}$	3·0$\frac{5}{8}$	4·6$\frac{7}{8}$	6·1$\frac{1}{8}$	7·7$\frac{1}{2}$	9·1$\frac{3}{4}$	10·8	12·2$\frac{3}{8}$	13·8$\frac{5}{8}$	15·2$\frac{7}{8}$
LENGTH OF HIP	1·9$\frac{7}{8}$	3·7$\frac{3}{4}$	5·5$\frac{5}{8}$	7·3$\frac{1}{2}$	9·1$\frac{3}{8}$	10·11$\frac{1}{4}$	12·9$\frac{1}{8}$	14·7	16·4$\frac{3}{4}$	18·2$\frac{5}{8}$

50° PITCH

RISE OF COMMON RAFTER 1′ 2$\frac{5}{16}$″ PER FOOT OF RUN

BEVELS:

1.	COMMON RAFTER	– SEAT	50
2.	″ ″	– RIDGE	40
3.	HIP OR VALLEY	– SEAT	40
4.	″ ″ ″	– RIDGE	50
5.	JACK RAFTER	– EDGE	32$\frac{1}{2}$
6.	PURLIN	– EDGE	57$\frac{1}{2}$
7.	″	– SIDE	52$\frac{1}{2}$

JACK RAFTERS 16 in. CENTRES DECREASE 2′ 0$\frac{7}{8}$″, 18 in.—2′ 4″, 24 in.—3′ 1$\frac{3}{8}$″

RUN OF RAFTER *ins.*	$\frac{1}{2}$	1	2	3	4	5	6	7	8	9	10	11
LENGTH OF RAFTER	$\frac{3}{4}$	1$\frac{1}{2}$	3$\frac{1}{8}$	4$\frac{5}{8}$	6$\frac{1}{4}$	7$\frac{3}{4}$	9$\frac{3}{8}$	10$\frac{7}{8}$	12$\frac{1}{2}$	14	15$\frac{1}{2}$	17$\frac{1}{8}$
LENGTH OF HIP	$\frac{7}{8}$	1$\frac{7}{8}$	3$\frac{3}{4}$	5$\frac{1}{2}$	7$\frac{3}{8}$	9$\frac{1}{4}$	11$\frac{1}{8}$	13	14$\frac{3}{4}$	16$\frac{5}{8}$	18$\frac{1}{2}$	20$\frac{3}{8}$

RUN OF RAFTER *ft.*	1	2	3	4	5	6	7	8	9	10
LENGTH OF RAFTER	1·6$\frac{5}{8}$	3·1$\frac{3}{8}$	4·8	6·2$\frac{5}{8}$	7·9$\frac{3}{8}$	9·4	10·10$\frac{5}{8}$	12·5$\frac{3}{8}$	14·0	15·6$\frac{5}{8}$
LENGTH OF HIP	1·10$\frac{1}{4}$	3·8$\frac{3}{8}$	5·6$\frac{5}{8}$	7·4$\frac{3}{4}$	9·3	11·1$\frac{1}{8}$	12·11$\frac{3}{8}$	14·9$\frac{1}{2}$	16·7$\frac{3}{4}$	18·5$\frac{7}{8}$

51° PITCH

RISE OF COMMON RAFTER 1′ 2$\frac{13}{16}$″ PER FOOT OF RUN

BEVELS:				
1.	COMMON RAFTER	–	SEAT	51
2.	″ ″	–	RIDGE	39
3.	HIP OR VALLEY	–	SEAT	41
4.	″ ″ ″	–	RIDGE	49
5.	JACK RAFTER	–	EDGE	32
6.	PURLIN	–	EDGE	58
7.	″	–	SIDE	52

JACK RAFTERS 16 in. CENTRES DECREASE 2′ 1$\frac{3}{8}$″, 18 in.—2′ 4$\frac{5}{8}$″, 24 in.—3′ 2$\frac{1}{8}$″

RUN OF RAFTER *ins.*	$\frac{1}{2}$	1	2	3	4	5	6	7	8	9	10	11
LENGTH OF RAFTER	$\frac{3}{4}$	1$\frac{5}{8}$	3$\frac{1}{8}$	4$\frac{3}{4}$	6$\frac{3}{8}$	8	9$\frac{1}{2}$	11$\frac{1}{8}$	12$\frac{3}{4}$	14$\frac{1}{4}$	15$\frac{7}{8}$	17$\frac{1}{2}$
LENGTH OF HIP	$\frac{7}{8}$	1$\frac{7}{8}$	3$\frac{3}{4}$	5$\frac{5}{8}$	7$\frac{1}{2}$	9$\frac{3}{8}$	11$\frac{1}{4}$	13$\frac{1}{8}$	15	16$\frac{7}{8}$	18$\frac{3}{4}$	20$\frac{5}{8}$

RUN OF RAFTER *ft.*	1	2	3	4	5	6	7	8	9	10
LENGTH OF RAFTER	1·7$\frac{1}{8}$	3·2$\frac{1}{8}$	4·9$\frac{1}{4}$	6·4$\frac{1}{4}$	7·11$\frac{3}{8}$	9·6$\frac{3}{8}$	11·1$\frac{1}{2}$	12·8$\frac{1}{2}$	14·3$\frac{5}{8}$	15·10$\frac{5}{8}$
LENGTH OF HIP	1·10$\frac{1}{2}$	3·9	5·7$\frac{5}{8}$	7·6$\frac{1}{8}$	9·4$\frac{5}{8}$	11·3$\frac{1}{4}$	13·1$\frac{3}{4}$	15·0$\frac{1}{4}$	16·10$\frac{7}{48}$	18·9$\frac{3}{8}$

52° PITCH

RISE OF COMMON RAFTER 1′ 3$\frac{3}{8}$″ PER FOOT OF RUN

BEVELS:
1. COMMON RAFTER – SEAT 52
2. " " – RIDGE 38
3. HIP OR VALLEY – SEAT 42
4. " " " – RIDGE 48
5. JACK RAFTER – EDGE 31$\frac{1}{2}$
6. PURLIN – EDGE 58$\frac{1}{2}$
7. " – SIDE 52

JACK RAFTERS 16 in. CENTRES DECREASE 2′ 2″, 18 in.—2′ 5$\frac{1}{4}$″, 24 in.—3′ 3″

RUN OF RAFTER *ins.*	$\frac{1}{2}$	1	2	3	4	5	6	7	8	9	10	11
LENGTH OF RAFTER	$\frac{3}{4}$	1$\frac{5}{8}$	3$\frac{1}{4}$	4$\frac{7}{8}$	6$\frac{1}{2}$	8$\frac{1}{8}$	9$\frac{3}{4}$	11$\frac{3}{8}$	13	14$\frac{5}{8}$	16$\frac{1}{4}$	17$\frac{7}{8}$
LENGTH OF HIP	1	1$\frac{7}{8}$	3$\frac{7}{8}$	5$\frac{3}{4}$	7$\frac{5}{8}$	9$\frac{1}{2}$	11$\frac{1}{2}$	13$\frac{3}{8}$	15$\frac{1}{4}$	17$\frac{1}{8}$	19$\frac{1}{8}$	21

RUN OF RAFTER *ft.*	1	2	3	4	5	6	7	8	9	10
LENGTH OF RAFTER	1·7$\frac{1}{2}$	3·3	4·10$\frac{1}{2}$	6·6	8·1$\frac{1}{2}$	9·9	11·4$\frac{1}{2}$	12·11$\frac{7}{8}$	14·7$\frac{3}{8}$	16·2$\frac{7}{8}$
LENGTH OF HIP	1·10$\frac{7}{8}$	3·9$\frac{3}{4}$	5·8$\frac{5}{8}$	7·7$\frac{1}{2}$	9·6$\frac{3}{8}$	11·5$\frac{1}{4}$	13·4$\frac{1}{4}$	15·3$\frac{1}{8}$	17·2	19·0$\frac{7}{8}$

53° PITCH

RISE OF COMMON RAFTER 1′ 3$\frac{15}{16}$″ PER FOOT OF RUN

BEVELS:				
1.	COMMON RAFTER	–	SEAT	53
2.	″ ″	–	RIDGE	37
3.	HIP OR VALLEY	–	SEAT	43
4.	″ ″ ″	–	RIDGE	47
5.	JACK RAFTER	–	EDGE	31
6.	PURLIN	–	EDGE	59
7.	″	–	SIDE	51$\frac{1}{2}$

JACK RAFTERS 16 in. CENTRES DECREASE 2′ 2$\frac{5}{8}$″, 18 in.—2′ 5$\frac{7}{8}$″, 24 in.—3′ 3$\frac{7}{8}$″

RUN OF RAFTER *ins.*	$\frac{1}{2}$	1	2	3	4	5	6	7	8	9	10	11
LENGTH OF RAFTER	$\frac{7}{8}$	1$\frac{5}{8}$	3$\frac{3}{8}$	5	6$\frac{5}{8}$	8$\frac{1}{4}$	10	11$\frac{5}{8}$	13$\frac{1}{4}$	15	16$\frac{5}{8}$	18$\frac{1}{4}$
LENGTH OF HIP	1	1$\frac{7}{8}$	3$\frac{7}{8}$	5$\frac{7}{8}$	7$\frac{3}{4}$	9$\frac{3}{4}$	11$\frac{5}{8}$	13$\frac{5}{8}$	15$\frac{1}{2}$	17$\frac{1}{2}$	19$\frac{3}{8}$	21$\frac{3}{8}$

RUN OF RAFTER *ft.*	1	2	3	4	5	6	7	8	9	10
LENGTH OF RAFTER	1·8	3·3$\frac{7}{8}$	4·11$\frac{7}{8}$	6·7$\frac{3}{4}$	8·3$\frac{3}{4}$	9·11$\frac{5}{8}$	11·7$\frac{5}{8}$	13·3$\frac{1}{2}$	14·11$\frac{1}{2}$	16·7$\frac{3}{8}$
LENGTH OF HIP	1·11$\frac{1}{4}$	3·10$\frac{1}{2}$	5·9$\frac{3}{4}$	7·9$\frac{1}{8}$	9·8$\frac{3}{8}$	11·7$\frac{5}{8}$	13·6$\frac{7}{8}$	15·6$\frac{1}{8}$	17·5$\frac{3}{8}$	19·4$\frac{5}{8}$

54° PITCH

RISE OF COMMON RAFTER 1′ 4½″ PER FOOT OF RUN

BEVELS:				
1.	COMMON RAFTER	–	SEAT	54
2.	″ ″	–	RIDGE	36
3.	HIP OR VALLEY	–	SEAT	44
4.	″ ″ ″	–	RIDGE	46
5.	JACK RAFTER	–	EDGE	30½
6.	PURLIN	–	EDGE	59½
7.	″	–	SIDE	51

JACK RAFTERS 16 in. CENTRES DECREASE 2′ 3″, 18 in.—2′ 6⅝″, 24 in.—3′ 4⅞″

RUN OF RAFTER *ins.*	½	1	2	3	4	5	6	7	8	9	10	11
LENGTH OF RAFTER	⅞	1¾	3⅜	5⅛	6¾	8½	10¼	11⅞	13⅝	15⅜	17	18¾
LENGTH OF HIP	1	2	4	5⅞	7⅞	9⅞	11⅞	13¾	15¾	17¾	19¾	21¾

RUN OF RAFTER *ft.*	1	2	3	4	5	6	7	8	9	10
LENGTH OF RAFTER	1·8⅜	3·4⅞	5·1¼	6·9⅝	8·6⅛	10·2½	11·10⅞	13·7¾	15·3¾	17·0⅛
LENGTH OF HIP	1·11⅝	3·11⅜	5·11	7·10¾	9·10⅜	11·10⅛	13·9¾	15·9½	17·9¼	19·8⅞

55° PITCH

RISE OF COMMON RAFTER $1' 5\frac{1}{8}''$ PER FOOT OF RUN

BEVELS:				
1.	COMMON RAFTER	–	SEAT	55
2.	″ ″	–	RIDGE	35
3.	HIP OR VALLEY	–	SEAT	$45\frac{1}{2}$
4.	″ ″ ″	–	RIDGE	$44\frac{1}{2}$
5.	JACK RAFTER	–	EDGE	30
6.	PURLIN	–	EDGE	60
7.	″	–	SIDE	$50\frac{1}{2}$

JACK RAFTERS 16 in. CENTRES DECREASE $2' 3\frac{7}{8}''$, 18 in.—$2' 7\frac{3}{8}''$, 24 in.—$3' 5\frac{7}{8}''$

RUN OF RAFTER *ins.*	$\frac{1}{2}$	1	2	3	4	5	6	7	8	9	10	11
LENGTH OF RAFTER	$\frac{7}{8}$	$1\frac{3}{4}$	$3\frac{1}{2}$	$5\frac{1}{4}$	7	$8\frac{3}{4}$	$10\frac{1}{2}$	$12\frac{1}{4}$	14	$15\frac{3}{4}$	$17\frac{3}{8}$	$19\frac{1}{8}$
LENGTH OF HIP	1	2	4	6	8	10	12	$14\frac{1}{8}$	$16\frac{1}{8}$	$18\frac{1}{8}$	$20\frac{1}{8}$	$22\frac{1}{8}$

RUN OF RAFTER *ft.*	1	2	3	4	5	6	7	8	9	10
LENGTH OF RAFTER	$1{\cdot}8\frac{7}{8}$	$3{\cdot}5\frac{7}{8}$	$5{\cdot}2\frac{3}{4}$	$6{\cdot}11\frac{5}{8}$	$8{\cdot}8\frac{5}{8}$	$10{\cdot}5\frac{1}{2}$	$12{\cdot}2\frac{1}{2}$	$13{\cdot}11\frac{3}{8}$	$15{\cdot}8\frac{1}{4}$	$17{\cdot}5\frac{1}{4}$
LENGTH OF HIP	$2{\cdot}0\frac{1}{8}$	$4{\cdot}0\frac{1}{4}$	$6{\cdot}0\frac{3}{8}$	$8{\cdot}0\frac{1}{2}$	$10{\cdot}0\frac{5}{8}$	$12{\cdot}0\frac{5}{8}$	$14{\cdot}0\frac{3}{4}$	$16{\cdot}0\frac{7}{8}$	18·1	$20{\cdot}1\frac{1}{8}$

Note: Italian Pitch (Rise = $\frac{3}{4}$ Span) has a rafter seat bevel of 56° 18′. Lengths, etc., are based on this angle.

ITALIAN PITCH

RISE OF COMMON RAFTER 1′ 6″ PER FOOT OF RUN

BEVELS:

1.	COMMON RAFTER	– SEAT	56$\frac{1}{2}$
2.	" "	– RIDGE	33$\frac{1}{2}$
3.	HIP OR VALLEY	– SEAT	46$\frac{1}{2}$
4.	" " "	– RIDGE	43$\frac{1}{2}$
5.	JACK RAFTER	– EDGE	29
6.	PURLIN	– EDGE	61
7.	"	– SIDE	50

JACK RAFTERS 16 in. CENTRES DECREASE 2′ 4$\frac{7}{8}$″, 18 in.—2′ 8$\frac{1}{2}$″, 24 in.—3′ 7$\frac{1}{4}$″

RUN OF RAFTER *ins.*	$\frac{1}{2}$	1	2	3	4	5	6	7	8	9	10	11
LENGTH OF RAFTER	$\frac{7}{8}$	1$\frac{3}{4}$	3$\frac{5}{8}$	5$\frac{3}{8}$	7$\frac{1}{4}$	9	10$\frac{3}{4}$	12$\frac{5}{8}$	14$\frac{3}{8}$	16$\frac{1}{4}$	18	19$\frac{7}{8}$
LENGTH OF HIP	1	2	4$\frac{1}{8}$	6$\frac{1}{8}$	8$\frac{1}{4}$	10$\frac{1}{4}$	12$\frac{3}{8}$	14$\frac{3}{8}$	16$\frac{1}{2}$	18$\frac{1}{2}$	20$\frac{5}{8}$	22$\frac{5}{8}$

RUN OF RAFTER *ft.*	1	2	3	4	5	6	7	8	9	10
LENGTH OF RAFTER	1·9$\frac{5}{8}$	3·7$\frac{1}{4}$	5·4$\frac{7}{8}$	7·2$\frac{1}{2}$	9·0$\frac{1}{8}$	10·9$\frac{3}{4}$	12·7$\frac{3}{4}$	14·5	16·2$\frac{5}{8}$	18·0$\frac{1}{4}$
LENGTH OF HIP	2·0$\frac{3}{4}$	4·1$\frac{1}{2}$	6·2$\frac{1}{4}$	8·3	10·3$\frac{5}{8}$	12·4$\frac{3}{8}$	14·5$\frac{1}{8}$	16·5$\frac{7}{8}$	18·6$\frac{5}{8}$	20·7$\frac{3}{8}$

57° PITCH

RISE OF COMMON RAFTER 1′ 6½″ PER FOOT OF RUN

BEVELS:				
	1.	COMMON RAFTER	– SEAT	57
	2.	″ ″	– RIDGE	33
	3.	HIP OR VALLEY	– SEAT	$47\frac{1}{2}$
	4.	″ ″ ″	– RIDGE	$42\frac{1}{2}$
	5.	JACK RAFTER	– EDGE	$28\frac{1}{2}$
	6.	PURLIN	– EDGE	$61\frac{1}{2}$
	7.	″	– SIDE	50

JACK RAFTERS 16 in. CENTRES DECREASE 2′ $5\frac{3}{8}$″, 18 in.—2′ 9″, 24 in.—3′ $8\frac{1}{8}$″

RUN OF RAFTER *ins.*	$\frac{1}{2}$	1	2	3	4	5	6	7	8	9	10	11
LENGTH OF RAFTER	$\frac{7}{8}$	$1\frac{7}{8}$	$3\frac{5}{8}$	$5\frac{1}{2}$	$7\frac{3}{8}$	$9\frac{1}{8}$	11	$12\frac{7}{8}$	$14\frac{3}{4}$	$16\frac{1}{2}$	$18\frac{3}{8}$	$20\frac{1}{4}$
LENGTH OF HIP	1	$2\frac{1}{8}$	$4\frac{1}{8}$	$6\frac{1}{4}$	$8\frac{3}{8}$	$10\frac{1}{2}$	$12\frac{1}{2}$	$14\frac{5}{8}$	$16\frac{3}{4}$	$18\frac{3}{4}$	$20\frac{7}{8}$	23

RUN OF RAFTER *ft.*	1	2	3	4	5	6	7	8	9	10
LENGTH OF RAFTER	1·10	3·$8\frac{1}{8}$	5·$6\frac{1}{8}$	7·$4\frac{1}{8}$	9·$2\frac{1}{8}$	11·$0\frac{1}{4}$	12·$10\frac{1}{4}$	14·$8\frac{1}{4}$	16·$6\frac{1}{4}$	18·$4\frac{3}{8}$
LENGTH OF HIP	2·$1\frac{1}{8}$	4·$2\frac{1}{8}$	6·$3\frac{1}{4}$	8·$4\frac{3}{8}$	10·$5\frac{3}{8}$	12·$6\frac{1}{2}$	14·$7\frac{5}{8}$	16·$8\frac{3}{4}$	18·$9\frac{3}{4}$	20·$10\frac{7}{8}$

58° PITCH

RISE OF COMMON RAFTER 1′ 7$\frac{3}{16}$″ PER FOOT OF RUN

BEVELS:				
	1.	COMMON RAFTER	– SEAT	58
	2.	″ ″	– RIDGE	32
	3.	HIP OR VALLEY	– SEAT	48$\frac{1}{2}$
	4.	″ ″ ″	– RIDGE	41$\frac{1}{2}$
	5.	JACK RAFTER	– EDGE	28
	6.	PURLIN	– EDGE	62
	7.	″	– SIDE	49$\frac{1}{2}$

JACK RAFTERS 16 in. CENTRES DECREASE 2′ 6$\frac{1}{4}$″, 18 in.—2′ 10″, 24 in.—3′ 9$\frac{1}{4}$″

RUN OF RAFTER *ins.*	$\frac{1}{2}$	1	2	3	4	5	6	7	8	9	10	11
LENGTH OF RAFTER	$\frac{7}{8}$	1$\frac{7}{8}$	3$\frac{3}{4}$	5$\frac{5}{8}$	7$\frac{1}{2}$	9$\frac{3}{8}$	11$\frac{3}{8}$	13$\frac{1}{4}$	15$\frac{1}{8}$	17	18$\frac{7}{8}$	20$\frac{3}{4}$
LENGTH OF HIP	1$\frac{1}{8}$	2$\frac{1}{8}$	4$\frac{1}{4}$	6$\frac{3}{8}$	8$\frac{1}{2}$	10$\frac{5}{8}$	12$\frac{3}{4}$	15	17$\frac{1}{8}$	19$\frac{1}{4}$	21$\frac{3}{8}$	23$\frac{1}{2}$

RUN OF RAFTER *ft.*	1	2	3	4	5	6	7	8	9	10
LENGTH OF RAFTER	1·10$\frac{5}{8}$	3·9$\frac{1}{4}$	5·7$\frac{7}{8}$	7·6$\frac{5}{8}$	9·5$\frac{1}{4}$	11·3$\frac{7}{8}$	13·2$\frac{1}{2}$	15·1$\frac{1}{8}$	16·11$\frac{3}{4}$	18·10$\frac{1}{2}$
LENGTH OF HIP	2·1$\frac{5}{8}$	4·3$\frac{1}{4}$	6·4$\frac{7}{8}$	8·6$\frac{1}{2}$	10·8$\frac{1}{8}$	12·9$\frac{3}{4}$	14·11$\frac{3}{8}$	17·1	19·2$\frac{5}{8}$	21·4$\frac{1}{4}$

59° PITCH

RISE OF COMMON RAFTER 1′ 8″ PER FOOT OF RUN

BEVELS:

1. COMMON RAFTER	–	SEAT	59
2. ″ ″	–	RIDGE	31
3. HIP OR VALLEY	–	SEAT	$49\frac{1}{2}$
4. ″ ″ ″	–	RIDGE	$40\frac{1}{2}$
5. JACK RAFTER	–	EDGE	$27\frac{1}{2}$
6. PURLIN	–	EDGE	$62\frac{1}{2}$
7. ″	–	SIDE	$49\frac{1}{2}$

JACK RAFTERS 16 in. CENTRES DECREASE 2′ $7\frac{1}{8}$″, 18 in.—2′ 11″, 24 in.—3′ $10\frac{5}{8}$″

RUN OF RAFTER *ins.*	$\frac{1}{2}$	1	2	3	4	5	6	7	8	9	10	11
LENGTH OF RAFTER	1	2	$3\frac{7}{8}$	$5\frac{7}{8}$	$7\frac{3}{4}$	$9\frac{3}{4}$	$11\frac{5}{8}$	$13\frac{5}{8}$	$15\frac{1}{2}$	$17\frac{1}{2}$	$19\frac{3}{8}$	$21\frac{3}{8}$
LENGTH OF HIP	$1\frac{1}{8}$	$2\frac{1}{4}$	$4\frac{3}{8}$	$6\frac{1}{2}$	$8\frac{3}{4}$	$10\frac{7}{8}$	$13\frac{1}{8}$	$15\frac{1}{4}$	$17\frac{1}{2}$	$19\frac{5}{8}$	$21\frac{7}{8}$	24

RUN OF RAFTER *ft.*	1	2	3	4	5	6	7	8	9	10
LENGTH OF RAFTER	$1{\cdot}11\frac{1}{4}$	$3{\cdot}10\frac{5}{8}$	$5{\cdot}9\frac{7}{8}$	$7{\cdot}9\frac{1}{4}$	$9{\cdot}8\frac{1}{2}$	$11{\cdot}7\frac{3}{4}$	$13{\cdot}7\frac{1}{8}$	$15{\cdot}6\frac{3}{8}$	$17{\cdot}5\frac{3}{4}$	19·5
LENGTH OF HIP	$2{\cdot}2\frac{1}{4}$	$4{\cdot}4\frac{3}{8}$	$6{\cdot}6\frac{5}{8}$	$8{\cdot}8\frac{7}{8}$	10·11	$13{\cdot}1\frac{1}{4}$	$15{\cdot}3\frac{1}{2}$	$17{\cdot}5\frac{5}{8}$	$19{\cdot}7\frac{7}{8}$	$21{\cdot}10\frac{1}{8}$

EQUILATERAL or 60° PITCH

RISE OF COMMON RAFTER 1′ $8\frac{13}{16}$″ PER FOOT OF RUN

BEVELS:				
1.	COMMON RAFTER	–	SEAT	60
2.	″ ″	–	RIDGE	30
3.	HIP OR VALLEY	–	SEAT	51
4.	″ ″ ″	–	RIDGE	39
5.	JACK RAFTER	–	EDGE	$26\frac{1}{2}$
6.	PURLIN	–	EDGE	$63\frac{1}{2}$
7.	″	–	SIDE	49

JACK RAFTERS 16 in. CENTRES DECREASE 2′ 8″, 18 in.—3′ 0″, 24 in.—4′ 0″

RUN OF RAFTER *ins.*	$\frac{1}{2}$	1	2	3	4	5	6	7	8	9	10	11
LENGTH OF RAFTER	1	2	4	6	8	10	12	14	16	18	20	22
LENGTH OF HIP	$1\frac{1}{8}$	$2\frac{1}{4}$	$4\frac{1}{2}$	$6\frac{3}{4}$	9	$11\frac{1}{8}$	$13\frac{3}{8}$	$15\frac{5}{8}$	$17\frac{7}{8}$	$20\frac{1}{8}$	$22\frac{3}{8}$	$24\frac{5}{8}$

RUN OF RAFTER *ft.*	1	2	3	4	5	6	7	8	9	10
LENGTH OF RAFTER	2·0	4·0	6·0	8·0	10·0	12·0	14·0	16·0	18·0	20·0
LENGTH OF HIP	2·$2\frac{7}{8}$	4·$5\frac{5}{8}$	6·$8\frac{1}{2}$	8·$11\frac{1}{4}$	11·$2\frac{1}{8}$	13·5	15·$7\frac{3}{4}$	17·$10\frac{5}{8}$	20·$1\frac{1}{2}$	22·$4\frac{3}{8}$

61° PITCH

RISE OF COMMON RAFTER 1′ $9\frac{5}{8}$″ PER FOOT OF RUN

BEVELS:

1.	COMMON RAFTER	–	SEAT	61
2.	″ ″	–	RIDGE	29
3.	HIP OR VALLEY	–	SEAT	52
4.	″ ″ ″	–	RIDGE	38
5.	JACK RAFTER	–	EDGE	26
6.	PURLIN	–	EDGE	64
7.	″	–	SIDE	49

JACK RAFTERS 16 in. CENTRES DECREASE 2′ 9″, 18 in.—3′ $1\frac{1}{8}$″, 24 in.—4′ $1\frac{1}{2}$″

RUN OF RAFTER *ins.*	$\frac{1}{2}$	1	2	3	4	5	6	7	8	9	10	11
LENGTH OF RAFTER	1	$2\frac{1}{8}$	$4\frac{1}{8}$	$6\frac{1}{4}$	$8\frac{1}{4}$	$10\frac{3}{8}$	$12\frac{3}{8}$	$14\frac{3}{8}$	$16\frac{1}{2}$	$18\frac{1}{2}$	$20\frac{5}{8}$	$22\frac{3}{4}$
LENGTH OF HIP	$1\frac{1}{8}$	$2\frac{1}{4}$	$4\frac{5}{8}$	$6\frac{7}{8}$	$9\frac{1}{8}$	$11\frac{1}{2}$	$13\frac{3}{4}$	16	$18\frac{3}{8}$	$20\frac{5}{8}$	$22\frac{7}{8}$	$25\frac{1}{4}$

RUN OF RAFTER *ft.*	1	2	3	4	5	6	7	8	9	10
LENGTH OF RAFTER	2·0$\frac{3}{4}$	4·1$\frac{1}{2}$	6·2$\frac{1}{4}$	8·3	10·3$\frac{3}{4}$	12·4$\frac{1}{2}$	14·5$\frac{1}{4}$	16·6	18·6$\frac{3}{4}$	20·7$\frac{1}{2}$
LENGTH OF HIP	2·3$\frac{1}{2}$	4·7	6·10$\frac{1}{2}$	9·2	11·5$\frac{1}{2}$	13·9	16·0$\frac{5}{8}$	18·4$\frac{1}{8}$	20·7$\frac{5}{8}$	22·11$\frac{1}{8}$

62° PITCH

RISE OF COMMON RAFTER 1′ 10$\frac{9}{16}$″ PER FOOT OF RUN

BEVELS:				
1.	COMMON RAFTER	–	SEAT	62
2.	″ ″	–	RIDGE	28
3.	HIP OR VALLEY	–	SEAT	53
4.	″ ″ ″	–	RIDGE	37
5.	JACK RAFTER	–	EDGE	25
6.	PURLIN	–	EDGE	65
7.	″	–	SIDE	48$\frac{1}{2}$

JACK RAFTERS 16 in. CENTRES DECREASE 2′ 10$\frac{1}{8}$″, 18 in.—3′ 2$\frac{3}{8}$″, 24 in.—4′ 3$\frac{1}{8}$″

RUN OF RAFTER *ins.*	$\frac{1}{2}$	1	2	3	4	5	6	7	8	9	10	11
LENGTH OF RAFTER	1$\frac{1}{8}$	2$\frac{1}{8}$	4$\frac{1}{4}$	6$\frac{3}{8}$	8$\frac{1}{2}$	10$\frac{5}{8}$	12$\frac{3}{4}$	14$\frac{7}{8}$	17	19$\frac{1}{8}$	21$\frac{1}{4}$	23$\frac{3}{8}$
LENGTH OF HIP	1$\frac{1}{8}$	2$\frac{3}{8}$	4$\frac{3}{4}$	7	9$\frac{3}{8}$	11$\frac{3}{4}$	14$\frac{1}{8}$	16$\frac{1}{2}$	18$\frac{7}{8}$	21$\frac{1}{8}$	23$\frac{1}{2}$	25$\frac{7}{8}$

RUN OF RAFTER *ft.*	1	2	3	4	5	6	7	8	9	10
LENGTH OF RAFTER	2·1$\frac{1}{2}$	4·3$\frac{1}{8}$	6·4$\frac{5}{8}$	8·6$\frac{1}{4}$	10·7$\frac{3}{4}$	12·9$\frac{3}{8}$	14·10$\frac{7}{8}$	17·0$\frac{1}{2}$	19·2	21·3$\frac{5}{8}$
LENGTH OF HIP	2·4$\frac{1}{4}$	4·8$\frac{1}{2}$	7·0$\frac{3}{4}$	9·5	11·9$\frac{1}{8}$	14·1$\frac{3}{8}$	16·5$\frac{5}{8}$	18·9$\frac{7}{8}$	21·2$\frac{1}{8}$	23·6$\frac{3}{8}$

Note: Gothic Pitch (Rise = Span) has a rafter seat bevel of 63° 26′. Lengths, etc., are based on this angle.

GOTHIC PITCH

RISE OF COMMON RAFTER 2′ 0″ PER FOOT OF RUN

BEVELS:				
1.	COMMON RAFTER	–	SEAT	$63\frac{1}{2}$
2.	″ ″	–	RIDGE	$26\frac{1}{2}$
3.	HIP OR VALLEY	–	SEAT	$54\frac{1}{2}$
4.	″ ″ ″	–	RIDGE	$35\frac{1}{2}$
5.	JACK RAFTER	–	EDGE	24
6.	PURLIN	–	EDGE	66
7.	″	–	SIDE	48

JACK RAFTERS 16 in. CENTRES DECREASE 2′ $11\frac{3}{4}$″, 18 in.—3′ $4\frac{1}{4}$″, 24 in.—4′ $5\frac{5}{8}$″

RUN OF RAFTER *ins.*	$\frac{1}{2}$	1	2	3	4	5	6	7	8	9	10	11
LENGTH OF RAFTER	$1\frac{1}{8}$	$2\frac{1}{4}$	$4\frac{1}{2}$	$6\frac{3}{4}$	9	$11\frac{1}{8}$	$13\frac{3}{8}$	$15\frac{5}{8}$	$17\frac{7}{8}$	$20\frac{1}{8}$	$22\frac{3}{8}$	$24\frac{5}{8}$
LENGTH OF HIP	$1\frac{1}{4}$	$2\frac{1}{2}$	$4\frac{7}{8}$	$7\frac{3}{8}$	$9\frac{3}{4}$	$12\frac{1}{4}$	$14\frac{3}{4}$	$17\frac{1}{8}$	$19\frac{5}{8}$	22	$24\frac{1}{2}$	27

RUN OF RAFTER *ft.*	1	2	3	4	5	6	7	8	9	10
LENGTH OF RAFTER	$2{\cdot}2\frac{7}{8}$	$4{\cdot}5\frac{5}{8}$	$6{\cdot}8\frac{1}{2}$	$8{\cdot}11\frac{3}{8}$	$11{\cdot}2\frac{1}{8}$	13·5	$15{\cdot}7\frac{7}{8}$	$17{\cdot}10\frac{5}{8}$	$20{\cdot}1\frac{1}{2}$	$22{\cdot}4\frac{1}{4}$
LENGTH OF HIP	$2{\cdot}5\frac{3}{8}$	$4{\cdot}10\frac{3}{4}$	$7{\cdot}4\frac{1}{8}$	$9{\cdot}9\frac{5}{8}$	12·3	$14{\cdot}8\frac{3}{8}$	$17{\cdot}1\frac{3}{4}$	$19{\cdot}7\frac{1}{8}$	$22{\cdot}0\frac{1}{2}$	24·6

65° PITCH

RISE OF COMMON RAFTER 2′ 1$\frac{3}{4}$″ PER FOOT OF RUN

BEVELS:				
	1. COMMON RAFTER	–	SEAT	65
	2. ″ ″	–	RIDGE	25
	3. HIP OR VALLEY	–	SEAT	56$\frac{1}{2}$
	4. ″ ″ ″	–	RIDGE	33$\frac{1}{2}$
	5. JACK RAFTER	–	EDGE	23
	6. PURLIN	–	EDGE	67
	7. ″	–	SIDE	48

JACK RAFTERS 16 in. CENTRES DECREASE 3′ 1$\frac{7}{8}$″, 18 in.—3′ 6$\frac{5}{8}$″, 24 in.—4′ 8$\frac{3}{4}$″

RUN OF RAFTER *ins.*	$\frac{1}{2}$	1	2	3	4	5	6	7	8	9	10	11
LENGTH OF RAFTER	1$\frac{1}{8}$	2$\frac{3}{8}$	4$\frac{3}{4}$	7$\frac{1}{8}$	9$\frac{1}{2}$	11$\frac{7}{8}$	14$\frac{1}{4}$	16$\frac{5}{8}$	18$\frac{7}{8}$	21$\frac{1}{4}$	23$\frac{5}{8}$	26
LENGTH OF HIP	1$\frac{1}{4}$	2$\frac{5}{8}$	5$\frac{1}{8}$	7$\frac{3}{4}$	10$\frac{1}{4}$	12$\frac{7}{8}$	15$\frac{3}{8}$	18	20$\frac{1}{2}$	23$\frac{1}{8}$	25$\frac{3}{4}$	28$\frac{1}{4}$

RUN OF RAFTER *ft.*	1	2	3	4	5	6	7	8	9	10
LENGTH OF RAFTER	2·4$\frac{3}{8}$	4·8$\frac{3}{4}$	7·1$\frac{1}{8}$	9·5$\frac{5}{8}$	11·10	14·2$\frac{3}{8}$	16·6$\frac{3}{4}$	18·11$\frac{1}{8}$	21·3$\frac{1}{2}$	23·8
LENGTH OF HIP	2·6$\frac{7}{8}$	5·1$\frac{5}{8}$	7·8$\frac{1}{2}$	10·3$\frac{1}{4}$	12·10$\frac{1}{8}$	15·5	17·11$\frac{3}{4}$	20·6$\frac{5}{8}$	23·1$\frac{3}{8}$	25·8$\frac{1}{4}$

66° PITCH

RISE OF COMMON RAFTER 2′ 2$\frac{15}{16}$″ PER FOOT OF RUN

BEVELS:				
1.	COMMON RAFTER	–	SEAT	66
2.	″ ″	–	RIDGE	24
3.	HIP OR VALLEY	–	SEAT	58
4.	″ ″ ″	–	RIDGE	32
5.	JACK RAFTER	–	EDGE	22
6.	PURLIN	–	EDGE	68
7.	″	–	SIDE	47$\frac{1}{2}$

JACK RAFTERS 16 in. CENTRES DECREASE 3′ 3$\frac{3}{8}$″, 18 in.—3′ 8$\frac{1}{4}$″, 24 in.—4′ 11″

RUN OF RAFTER *ins.*	$\frac{1}{2}$	1	2	3	4	5	6	7	8	9	10	11
LENGTH OF RAFTER	1$\frac{1}{4}$	2$\frac{1}{2}$	4$\frac{7}{8}$	7$\frac{3}{8}$	9$\frac{7}{8}$	12$\frac{1}{4}$	14$\frac{3}{4}$	17$\frac{1}{4}$	19$\frac{5}{8}$	22$\frac{1}{8}$	24$\frac{5}{8}$	27
LENGTH OF HIP	1$\frac{3}{8}$	2$\frac{5}{8}$	5$\frac{1}{4}$	8	10$\frac{5}{8}$	13$\frac{1}{4}$	15$\frac{7}{8}$	18$\frac{5}{8}$	21$\frac{1}{4}$	23$\frac{7}{8}$	26$\frac{1}{2}$	29$\frac{1}{4}$

RUN OF RAFTER *ft.*	1	2	3	4	5	6	7	8	9	10
LENGTH OF RAFTER	2·5$\frac{1}{2}$	4·11	7·4$\frac{1}{2}$	9·10	12·3$\frac{1}{2}$	14·9	17·2$\frac{1}{2}$	19·8	22·1$\frac{1}{2}$	24·7
LENGTH OF HIP	2·7$\frac{7}{8}$	5·3$\frac{3}{4}$	7·11$\frac{1}{2}$	10·7$\frac{3}{8}$	13·3$\frac{1}{4}$	15·11	18·6$\frac{7}{8}$	21·2$\frac{3}{4}$	23·10$\frac{5}{8}$	26·6$\frac{3}{8}$

67° PITCH

RISE OF COMMON RAFTER 2′ 4$\frac{1}{4}$″ PER FOOT OF RUN

BEVELS:				
	1.	COMMON RAFTER	– SEAT	67
	2.	″ ″	– RIDGE	23
	3.	HIP OR VALLEY	– SEAT	59
	4.	″ ″ ″	– RIDGE	31
	5.	JACK RAFTER	– EDGE	21$\frac{1}{2}$
	6.	PURLIN	– EDGE	68$\frac{1}{2}$
	7.	″	– SIDE	47$\frac{1}{2}$

JACK RAFTERS 16 in. CENTRES DECREASE 3′ 5″, 18 in.—3′ 10$\frac{1}{8}$″, 24 in.—5′ 1$\frac{1}{2}$″

RUN OF RAFTER *ins.*	$\frac{1}{2}$	1	2	3	4	5	6	7	8	9	10	11
LENGTH OF RAFTER	1$\frac{1}{4}$	2$\frac{1}{2}$	5$\frac{1}{8}$	7$\frac{5}{8}$	10$\frac{1}{4}$	12$\frac{3}{4}$	15$\frac{3}{8}$	17$\frac{7}{8}$	20$\frac{1}{2}$	23	25$\frac{5}{8}$	28$\frac{1}{8}$
LENGTH OF HIP	1$\frac{3}{8}$	2$\frac{3}{4}$	5$\frac{1}{2}$	8$\frac{1}{4}$	11	13$\frac{3}{4}$	16$\frac{1}{2}$	19$\frac{1}{4}$	22	24$\frac{3}{4}$	27$\frac{1}{2}$	30$\frac{1}{4}$

RUN OF RAFTER *ft.*	1	2	3	4	5	6	7	8	9	10
LENGTH OF RAFTER	2·6$\frac{3}{4}$	5·1$\frac{3}{8}$	7·8$\frac{1}{8}$	10·2$\frac{7}{8}$	12·9$\frac{1}{2}$	15·4$\frac{1}{4}$	17·11	20·5$\frac{3}{4}$	23·0$\frac{3}{8}$	25·7$\frac{1}{8}$
LENGTH OF HIP	2·9	5·5$\frac{7}{8}$	8·2$\frac{7}{8}$	10·11$\frac{7}{8}$	13·8$\frac{7}{8}$	16·5$\frac{3}{4}$	19·2$\frac{3}{4}$	21·11$\frac{3}{4}$	24·8$\frac{3}{4}$	27·5$\frac{5}{8}$

68° PITCH

RISE OF COMMON RAFTER 2′ 5$\frac{11}{16}$″ PER FOOT OF RUN

BEVELS:				
1.	COMMON RAFTER	–	SEAT	68
2.	″ ″	–	RIDGE	22
3.	HIP OR VALLEY	–	SEAT	60$\frac{1}{2}$
4.	″ ″ ″	–	RIDGE	29$\frac{1}{2}$
5.	JACK RAFTER	–	EDGE	20$\frac{1}{2}$
6.	PURLIN	–	EDGE	69$\frac{1}{2}$
7.	″	–	SIDE	47

JACK RAFTERS 16 in. CENTRES DECREASE 3′ 6$\frac{3}{4}$″, 18 in.—4′ 0″, 24 in.—5′ 4$\frac{1}{8}$″

RUN OF RAFTER *ins.*	$\frac{1}{2}$	1	2	3	4	5	6	7	8	9	10	11
LENGTH OF RAFTER	1$\frac{3}{8}$	2$\frac{5}{8}$	5$\frac{3}{8}$	8	10$\frac{5}{8}$	13$\frac{3}{8}$	16	18$\frac{5}{8}$	21$\frac{3}{8}$	24	26$\frac{3}{4}$	29$\frac{3}{8}$
LENGTH OF HIP	1$\frac{3}{8}$	2$\frac{7}{8}$	5$\frac{3}{4}$	8$\frac{1}{2}$	11$\frac{3}{8}$	14$\frac{1}{4}$	17$\frac{1}{8}$	20	22$\frac{3}{4}$	25$\frac{5}{8}$	28$\frac{1}{2}$	31$\frac{3}{8}$

RUN OF RAFTER *ft.*	1	2	3	4	5	6	7	8	9	10
LENGTH OF RAFTER	2·8	5·4$\frac{1}{8}$	8·0$\frac{1}{8}$	10·8$\frac{1}{8}$	13·4$\frac{1}{8}$	16·0$\frac{1}{4}$	18·8$\frac{1}{4}$	21·4$\frac{1}{4}$	24·0$\frac{1}{4}$	26·8$\frac{3}{8}$
LENGTH OF HIP	2·10$\frac{1}{4}$	5·8$\frac{3}{8}$	8·6$\frac{5}{8}$	11·4$\frac{7}{8}$	14·3	17·1$\frac{1}{4}$	19·11$\frac{1}{2}$	22·9$\frac{5}{8}$	25·7$\frac{7}{8}$	28·6$\frac{1}{8}$

69° PITCH

RISE OF COMMON RAFTER 2′ 7¼″ PER FOOT OF RUN

BEVELS:			
1.	COMMON RAFTER	– SEAT	69
2.	″ ″	– RIDGE	21
3.	HIP OR VALLEY	– SEAT	61½
4.	″ ″ ″	– RIDGE	28½
5.	JACK RAFTER	– EDGE	19½
6.	PURLIN	– EDGE	70½
7.	″	– SIDE	47

JACK RAFTERS 16 in. CENTRES DECREASE 3′ 8⅝″, 18 in.—4′ 2¼″, 24 in.—5′ 7″

RUN OF RAFTER *ins.*	½	1	2	3	4	5	6	7	8	9	10	11
LENGTH OF RAFTER	1⅜	2¾	5⅜	8⅜	11⅛	14	16¾	19½	22⅜	25⅛	27⅞	30¾
LENGTH OF HIP	1½	3	5⅞	8⅞	11⅞	14⅞	17¾	20¾	23¾	26⅝	29⅝	32⅝

RUN OF RAFTER *ft.*	1	2	3	4	5	6	7	8	9	10
LENGTH OF RAFTER	2·9½	5·7	8·4¼	11·2	13·11⅜	16·8⅞	19·6⅜	22·3⅞	25·1⅜	27·10⅞
LENGTH OF HIP	2·11⅝	5·11⅛	8·10¾	11·10¼	14·9⅞	17·9⅜	20·9	23·8½	26·8⅛	29·7⅝

70° PITCH

RISE OF COMMON RAFTER 2′ 9″ PER FOOT OF RUN

BEVELS:				
1.	COMMON RAFTER	–	SEAT	70
2.	″ ″	–	RIDGE	20
3.	HIP OR VALLEY	–	SEAT	63
4.	″ ″ ″	–	RIDGE	27
5.	JACK RAFTER	–	EDGE	19
6.	PURLIN	–	EDGE	71
7.	″	–	SIDE	47

JACK RAFTERS 16 in. CENTRES DECREASE 3′ $10\frac{3}{4}$″, 18 in.—4′ $4\frac{5}{8}$″, 24 in.—5′ $10\frac{1}{8}$″

RUN OF RAFTER *ins.*	$\frac{1}{2}$	1	2	3	4	5	6	7	8	9	10	11
LENGTH OF RAFTER	$1\frac{1}{2}$	$2\frac{7}{8}$	$5\frac{7}{8}$	$8\frac{3}{4}$	$11\frac{3}{4}$	$14\frac{5}{8}$	$17\frac{1}{2}$	$20\frac{1}{2}$	$23\frac{3}{8}$	$26\frac{3}{8}$	$29\frac{1}{4}$	$32\frac{1}{8}$
LENGTH OF HIP	$1\frac{1}{2}$	$3\frac{1}{8}$	$6\frac{1}{8}$	$9\frac{1}{4}$	$12\frac{3}{8}$	$15\frac{1}{2}$	$18\frac{1}{2}$	$21\frac{5}{8}$	$24\frac{3}{4}$	$27\frac{3}{4}$	$30\frac{7}{8}$	34

RUN OF RAFTER *ft.*	1	2	3	4	5	6	7	8	9	10
LENGTH OF RAFTER	$2{\cdot}11\frac{1}{8}$	$5{\cdot}10\frac{1}{8}$	$8{\cdot}9\frac{1}{4}$	$11{\cdot}8\frac{3}{8}$	$14{\cdot}7\frac{3}{8}$	$17{\cdot}6\frac{1}{2}$	$20{\cdot}5\frac{5}{8}$	$23{\cdot}4\frac{5}{8}$	$26{\cdot}3\frac{3}{4}$	$29{\cdot}2\frac{7}{8}$
LENGTH OF HIP	$3{\cdot}1\frac{1}{8}$	$6{\cdot}2\frac{1}{8}$	$9{\cdot}3\frac{1}{4}$	$12{\cdot}4\frac{3}{8}$	$15{\cdot}5\frac{3}{8}$	$18{\cdot}6\frac{1}{2}$	$21{\cdot}7\frac{1}{2}$	$24{\cdot}8\frac{5}{8}$	$27{\cdot}9\frac{3}{4}$	$30{\cdot}10\frac{3}{4}$

71° PITCH

RISE OF COMMON RAFTER 2′ $10\frac{7}{8}$″ PER FOOT OF RUN

BEVELS:				
1.	COMMON RAFTER	–	SEAT	71
2.	″ ″	–	RIDGE	19
3.	HIP OR VALLEY	–	SEAT	64
4.	″ ″ ″	–	RIDGE	26
5.	JACK RAFTER	–	EDGE	18
6.	PURLIN	–	EDGE	72
7.	″	–	SIDE	$46\frac{1}{2}$

JACK RAFTERS 16 in. CENTRES DECREASE 4′ $1\frac{1}{8}$″, 18 in.—4′ $7\frac{1}{4}$″, 24 in.—6′ $1\frac{3}{4}$″

RUN OF RAFTER *ins.*	$\frac{1}{2}$	1	2	3	4	5	6	7	8	9	10	11
LENGTH OF RAFTER	$1\frac{1}{2}$	$3\frac{1}{8}$	$6\frac{1}{8}$	$9\frac{1}{4}$	$12\frac{1}{4}$	$15\frac{3}{8}$	$18\frac{3}{8}$	$21\frac{1}{2}$	$24\frac{5}{8}$	$27\frac{5}{8}$	$30\frac{3}{4}$	$33\frac{3}{4}$
LENGTH OF HIP	$1\frac{5}{8}$	$3\frac{1}{4}$	$6\frac{1}{2}$	$9\frac{3}{4}$	$12\frac{7}{8}$	$16\frac{1}{8}$	$19\frac{3}{8}$	$22\frac{5}{8}$	$25\frac{7}{8}$	$29\frac{1}{8}$	$32\frac{1}{4}$	$35\frac{1}{2}$

RUN OF RAFTER *ft.*	1	2	3	4	5	6	7	8	9	10
LENGTH OF RAFTER	$3{\cdot}0\frac{7}{8}$	$6{\cdot}1\frac{3}{4}$	$9{\cdot}2\frac{5}{8}$	$12{\cdot}3\frac{3}{8}$	$15{\cdot}4\frac{1}{4}$	$18{\cdot}5\frac{1}{8}$	21·6	$24{\cdot}6\frac{7}{8}$	$27{\cdot}7\frac{3}{4}$	$30{\cdot}8\frac{5}{8}$
LENGTH OF HIP	$3{\cdot}2\frac{3}{4}$	$6{\cdot}5\frac{1}{2}$	$9{\cdot}8\frac{1}{4}$	12·11	$16{\cdot}1\frac{3}{4}$	$19{\cdot}4\frac{1}{2}$	$22{\cdot}7\frac{3}{8}$	$25{\cdot}10\frac{1}{8}$	$29{\cdot}0\frac{7}{8}$	$32{\cdot}3\frac{5}{8}$

72° PITCH

RISE OF COMMON RAFTER 3′ 0$\frac{15}{16}$″ PER FOOT OF RUN

BEVELS:				
1.	COMMON RAFTER	–	SEAT	72
2.	" "	–	RIDGE	18
3.	HIP OR VALLEY	–	SEAT	65$\frac{1}{2}$
4.	" " "	–	RIDGE	24$\frac{1}{2}$
5.	JACK RAFTER	–	EDGE	17
6.	PURLIN	–	EDGE	73
7.	"	–	SIDE	46$\frac{1}{2}$

JACK RAFTERS 16 in. CENTRES DECREASE 4′ 3$\frac{3}{4}$″, 18 in.—4′ 10$\frac{1}{4}$″, 24 in.—6′ 5$\frac{5}{8}$″

RUN OF RAFTER *ins.*	$\frac{1}{2}$	1	2	3	4	5	6	7	8	9	10	11
LENGTH OF RAFTER	1$\frac{5}{8}$	3$\frac{1}{4}$	6$\frac{1}{2}$	9$\frac{3}{4}$	13	16$\frac{1}{8}$	19$\frac{3}{8}$	22$\frac{5}{8}$	25$\frac{7}{8}$	29$\frac{1}{8}$	32$\frac{3}{8}$	35$\frac{5}{8}$
LENGTH OF HIP	1$\frac{3}{4}$	3$\frac{3}{8}$	6$\frac{3}{4}$	10$\frac{1}{8}$	13$\frac{1}{2}$	16$\frac{7}{8}$	20$\frac{3}{8}$	23$\frac{3}{4}$	27$\frac{1}{8}$	30$\frac{1}{2}$	33$\frac{7}{8}$	37$\frac{1}{4}$

RUN OF RAFTER *ft.*	1	2	3	4	5	6	7	8	9	10
LENGTH OF RAFTER	3·2$\frac{7}{8}$	6·5$\frac{5}{8}$	9·8$\frac{1}{2}$	12·11$\frac{3}{8}$	16·2$\frac{1}{8}$	19·5	22·7$\frac{7}{8}$	25·10$\frac{5}{8}$	29·1$\frac{1}{2}$	32·4$\frac{3}{8}$
LENGTH OF HIP	3·4$\frac{5}{8}$	6·9$\frac{1}{4}$	10·1$\frac{7}{8}$	13·6$\frac{1}{2}$	16·11$\frac{1}{8}$	20·3$\frac{3}{4}$	23·8$\frac{3}{8}$	27·1	30·5$\frac{3}{4}$	33·10$\frac{3}{8}$

73° PITCH

RISE OF COMMON RAFTER 3′ 3¼″ PER FOOT OF RUN

BEVELS:				
1.	COMMON RAFTER	–	SEAT	73
2.	″ ″	–	RIDGE	17
3.	HIP OR VALLEY	–	SEAT	66½
4.	″ ″ ″	–	RIDGE	23½
5.	JACK RAFTER	–	EDGE	16½
6.	PURLIN	–	EDGE	73½
7.	″	–	SIDE	46½

JACK RAFTERS 16 in. CENTRES DECREASE 4′ 6¾″, 18 in.—5′ 1½″, 24 in.—6′ 10⅛″

RUN OF RAFTER *ins.*	½	1	2	3	4	5	6	7	8	9	10	11
LENGTH OF RAFTER	1¾	3⅜	6⅞	10¼	13⅝	17⅛	20½	24	27⅜	30¾	34¼	37⅝
LENGTH OF HIP	1¾	3⅝	7⅛	10¾	14¼	17⅞	21⅜	25	28⅛	32⅛	35⅝	39¼

RUN OF RAFTER *ft.*	1	2	3	4	5	6	7	8	9	10
LENGTH OF RAFTER	3·5	6·10⅛	10·3⅛	13·8⅛	17·1¼	20·6¼	23·11¼	27·4⅜	30·9⅜	34·2⅜
LENGTH OF HIP	3·6¾	7·1½	10·8¼	14·3	17·9¾	21·4½	24·11¼	28·6	32·0¾	35·7½

74° PITCH

RISE OF COMMON RAFTER 3′ 5⅞″ PER FOOT OF RUN

BEVELS:				
1.	COMMON RAFTER	–	SEAT	74
2.	″ ″	–	RIDGE	16
3.	HIP OR VALLEY	–	SEAT	68
4.	″ ″ ″	–	RIDGE	22
5.	JACK RAFTER	–	EDGE	15½
6.	PURLIN	–	EDGE	74½
7.	″	–	SIDE	46

JACK RAFTERS 16 in. CENTRES DECREASE 4′ 10″, 18 in.—5′ 5¼″, 24 in.—7′ 3⅛″

RUN OF RAFTER *ins.*	½	1	2	3	4	5	6	7	8	9	10	11
LENGTH OF RAFTER	1⅞	3⅝	7¼	10⅞	14½	18⅛	21¾	25⅜	29	32⅝	36¼	39⅞
LENGTH OF HIP	1⅞	3¾	7½	11¼	15	18¾	22⅝	26⅜	30⅛	33⅝	37⅝	41⅜

RUN OF RAFTER *ft.*	1	2	3	4	5	6	7	8	9	10
LENGTH OF RAFTER	3·7½	7·3⅛	10·10⅝	14·6⅛	18·1⅝	21·9¼				
LENGTH OF HIP	3·9⅛	7·6¼	11·3½	15·0⅝	18·9¾	22·7				

75° PITCH

> **Note**—Pitches steeper than 75° are not likely to be met with by the craftsman except in steeples, spires and turrets. These call for special construction and are not within the scope of this ready reckoner.

RISE OF COMMON RAFTER 3′ 8$\frac{13}{16}$″ PER FOOT OF RUN

BEVELS:				
1.	COMMON RAFTER	–	SEAT	75
2.	″ ″	–	RIDGE	15
3.	HIP OR VALLEY	–	SEAT	69
4.	″ ″ ″	–	RIDGE	21
5.	JACK RAFTER	–	EDGE	14$\frac{1}{2}$
6.	PURLIN	–	EDGE	75$\frac{1}{2}$
7.	″	–	SIDE	46

JACK RAFTERS 16 in. CENTRES DECREASE 5′ 1$\frac{7}{8}$″, 18 in.—5′ 9$\frac{1}{2}$″

RUN OF RAFTER *ins.*	$\frac{1}{2}$	1	2	3	4	5	6	7	8	9	10	11
LENGTH OF RAFTER	1$\frac{7}{8}$	3$\frac{7}{8}$	7$\frac{3}{4}$	11$\frac{5}{8}$	15$\frac{1}{2}$	19$\frac{3}{8}$	23$\frac{1}{8}$	27	30$\frac{7}{8}$	34$\frac{3}{4}$	38$\frac{5}{8}$	42$\frac{1}{2}$
LENGTH OF HIP	2	4	8	12	16	20	24	27$\frac{7}{8}$	31$\frac{7}{8}$	35$\frac{7}{8}$	39$\frac{7}{8}$	43$\frac{7}{8}$

RUN OF RAFTER *ft.*	1	2	3	4	5	6	7	8	9	10
LENGTH OF RAFTER	3·10$\frac{3}{8}$	7·8$\frac{3}{4}$	11·7$\frac{1}{8}$	15·5$\frac{1}{2}$	19·3$\frac{7}{8}$	23·2$\frac{1}{8}$				
LENGTH OF HIP	3·11$\frac{7}{8}$	7·11$\frac{3}{4}$	11·11$\frac{5}{8}$	15·11$\frac{1}{2}$	19·11$\frac{3}{8}$	23·11$\frac{3}{8}$				

6 WALL PLATE AND GABLE STRAPPING

Earlier in this book it was stated that the wall plate is the foundation to the roof, and like all foundations needs to be sound and secure. The wall plate should be strapped down to the building structure; this is usually done by using steel straps as illustrated in Fig. 12(a). The straps can either be built into the brick wall below, or face-fixed by screwing into the wall. On timber framed housing the plate may be adequately secured to the frame by nailing at centres specified by the designer.

Gable walls, especially those on steep pitched roof constructions where the gable is very tall, rely on the roof for their stability and NOT the other way around. Wind blowing on one gable exerts a pressure on it pushing it into the roof, whilst at the opposite end it creates a suction which attempts to suck the gable from the roof. It is therefore a requirement of the building regulations that the gables must be adequately tied back into the roof to give them support. Fig. 12(b) illustrates a typical gable end restraint system on a trussed rafter roof, but this equally applies to traditional roof construction. Building the purlins into the gable will help, but will only be effective if the purlin is mechanically fixed to the wall with some additional form of cleat or strap. Straps are normally placed at approximately 2 m centres, and it is essential that the strap is supported by solid blocking beneath it to ensure that it does not buckle, and that the last rafter is solidly blocked to the gable wall itself.

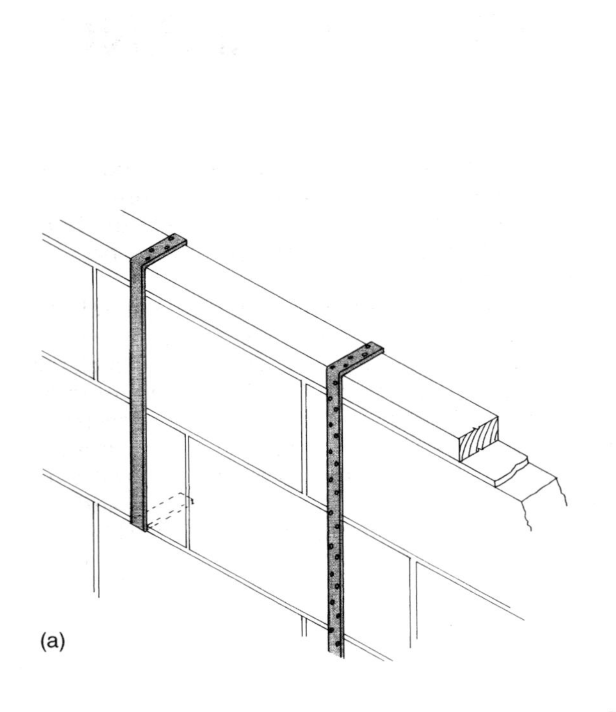
(a)

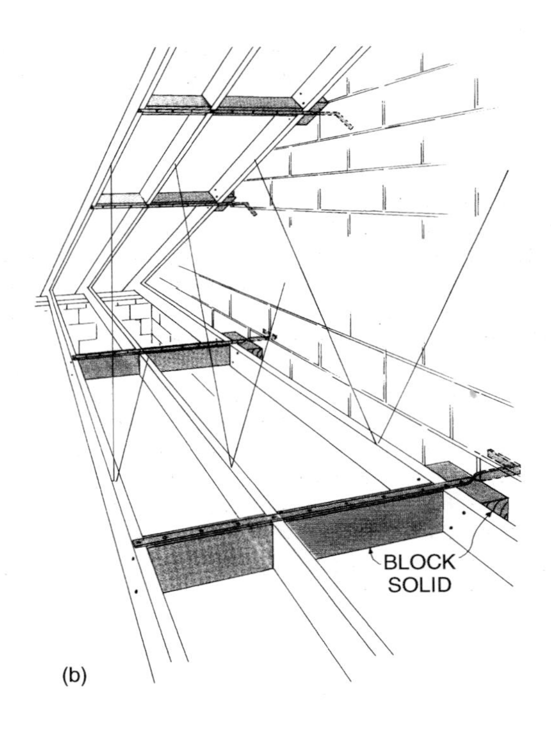

(b)

Fig. 12

7 WIND BRACING

From the previous chapter it can be seen that wind plays a considerable part in destabilising a structure and measures must be taken to ensure the stability of the roof construction in high wind situations. From the previous chapter strapping the gables to the roof have ensured their integrity with the roof, but apart from the binders, purlins and ridge, which connect the rafter members on a horizontal plane, the roof structure is still no more than a number of vertical members of timber connected with a limited number of nails to the members mentioned above. The roof is in effect no more than a set of dominoes standing vertically on their ends and which can be easily made to fall if a pressure is applied to the last one in line, in the case of a roof this would be the gable end. To prevent the domino toppling effect, wind bracing is introduced into the structure to triangulate it on a vertical plane. The elements being discussed here are set out in Fig. 13, which details the bracing required for a typical trussed rafter roof, much of this being applicable to a traditionally constructed roof without hips. Even on a hip roof where the ridge is twice as long as the length on plan of the hip itself it is wise to introduce wind bracing.

Wind bracing is usually timber typically of 25 × 100 mm (1″ × 4″) in cross section, fitted from wall plate to the ridge at an angle of approximately 45° on the underside of the rafters. At each rafter crossing the wind bracing should be nailed to the rafter with 3 nails. The braces should be fitted from the foot of the gable to the ridge on both sides of the roof and then at 45° back down again to the plate over the entire length of the roof. This is brace F in Fig. 13; brace H may well be replaced by a purlin in a traditionally cut roof, and brace G by the ridge board. Brace G at ceiling joist level, may well be ceiling joist binders on a traditionally cut roof, whilst K should be fitted to all types of roof construction. Brace J applies only to trussed rafter roof construction.

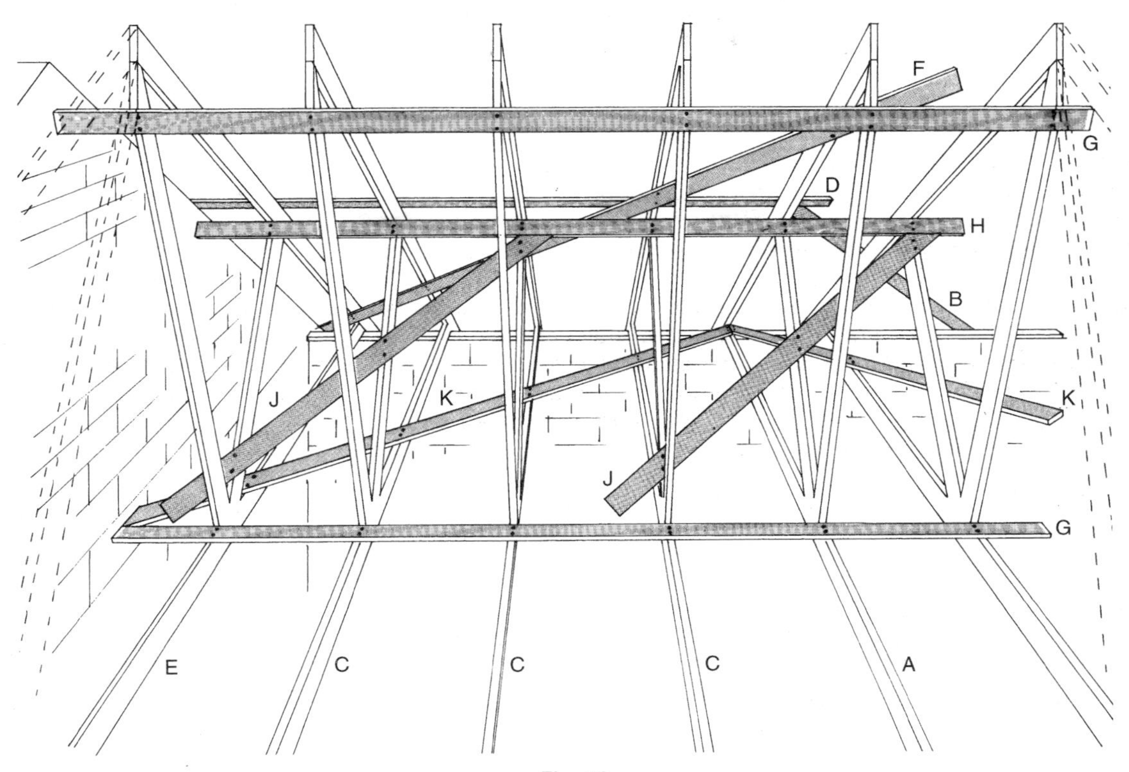

Fig. 13

8 ROOFING METALWORK

The use of smaller timber sections particularly in engineered roofs such as trussed rafter prefabricated assemblies, has led to the increased use of metalwork to join the various members together. Traditional 'tosh' or 'skew' nailing can easily split these smaller timber sections resulting in a poor connection. Fig. 14 illustrates some items of roofing metalwork.

All metalwork used in roof construction should be galvanised; this includes the nails to prevent corrosion. The modern roof can suffer from localised condensation especially if not adequately ventilated and this could lead to premature corrosion of nails and fixings. Whilst a galvanised nail has a slightly rough surface compared to a plain wire nail and therefore gives an improved resistance to movement in the joint, many of the metal to timber connections are specified to be fixed with square twisted galvanised nails which give a far improved performance in the joint.

Typical roofing metalwork would be as follows:

(a) Wall plate straps cross section 30 mm × 1.5/2 mm
(b) Gable restraint straps 30 mm × 5 mm.
(c) Trussed rafter clips to hold truss or rafter to wall plate.
(d) Hip corner tie to hold hip to wall plate at corner.
(e) Girder truss shoes to carry trussed rafters on girder truss support points.
(f) Multi nail plates for coupling timbers into longer lengths.
(g) Framing anchors to connect various trimmed openings and elements in the ceiling joist structure.

Care must be taken to use the nails specified by the metalwork manufacturer both in the type of nail to be used and the number of nails to be used in each joint.

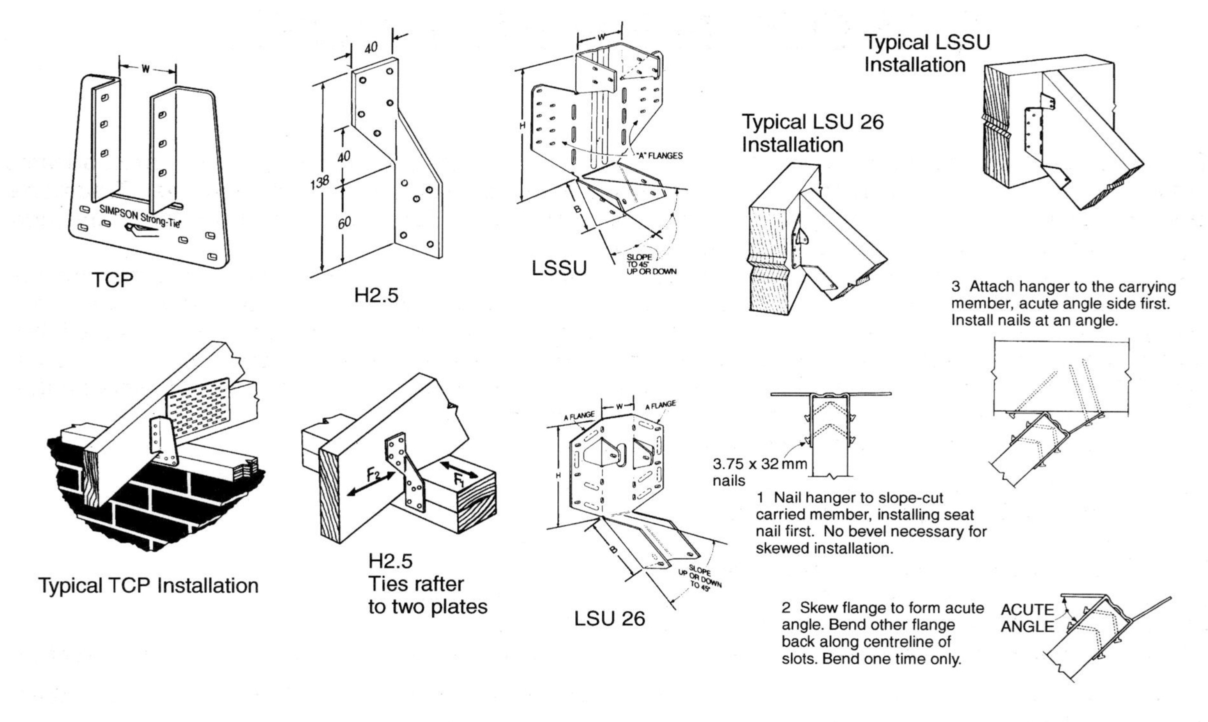
W
SIMPSON Strong-Tie
TCP
40
40
138
60
H2.5
W
H
"A" FLANGES
B
SLOPE TO 45° UP OR DOWN
LSSU
Typical LSSU Installation
Typical LSU 26 Installation
3 Attach hanger to the carrying member, acute angle side first. Install nails at an angle.
Typical TCP Installation
F2
F1
H2.5
Ties rafter to two plates
A FLANGE
W
A FLANGE
H
B
SLOPE UP OR DOWN TO 45°
LSU 26
3.75 x 32 mm nails
1 Nail hanger to slope-cut carried member, installing seat nail first. No bevel necessary for skewed installation.
2 Skew flange to form acute angle. Bend other flange back along centreline of slots. Bend one time only.
ACUTE ANGLE

Fig. 14

9 TOOLS AND EQUIPMENT

The roofing carpenter will need a number of tools and pieces of equipment to satisfactorily obtain information from drawings, mark the timber, set out the roof on the wall plate, cut the timber, and check the completed roof for line, level and plumb. A conventional pencil or pen may be needed for paper calculations, but the true carpenter's pencil should be used for marking timber.

Obtaining Information from the Drawing

(a) Scale rule: only to be used if dimensions are not clearly shown on the drawing.
(b) Protractor: to measure the angle of the roof, but again only if the angle is not written on the drawing.
(c) To mark out the length and angles to be cut on the timbers: steel measuring tape of minimum 5 m in length now available with a digital display to remove the possible error of misreading.
(d) A bevel: a simple carpenter's adjustable bevel is adequate, this being set to the protractor to obtain cutting angles.
(e) Alternatively a combination square with centre head and built-in protractor is more versatile.
(f) Alternatively a digital bevel may be used for instant visual display of the setting angles required.
(g) A traditional roofing square can be used if the carpenter is trained in the use of this particular tool.

To Cut the Roof

(h) Hand saw.
(i) Mains or 110 volt electric hand saw. Cordless powered hand saws are available, but a mains power source is still required to recharge batteries.

(j) A compound angle mitre saw. This is an electrically powered saw designed especially to cut angles on timbers – check that the saw is large enough to cope with the length of the cut required, a 300 mm diameter saw should be adequate. This type of saw can tilt in both planes and therefore be set to cut compound angles i.e. the ridge and edge bevels on hip jack rafters, in one operation. This type of saw is invariably fixed to a bench or stand, and therefore support will be needed for long timbers to be cut. This support should be fitted with a sliding stop system to allow quick repeat lengths to be cut without remeasuring. NB for safety all power tools on site should be 110 volts – for home use a 240 volt saw may be used only in conjunction with a power protection plug adapter – this will protect against accidentally cut cables and faulty wiring possibly causing electric shock to the operator.

Setting-Up the Roof Structure

(k) A steel tape at least as long as the roof wall plate itself.
(l) A good level, by this we mean a good quality level at least 900 mm long with two spirit levels one for horizontal use and one for vertical use for plumbing timbers.
(m) Alternatively a good quality level as above but with digital angle readout would be useful to check work on the roof as it proceeds.
(n) To check and set level and verticals over longer distances (i.e. beyond the 900 mm of the levels itself), a laser level could be used. A wide range of laser beam projecting levels are now available giving an accurate beam projection of up to 50 m. The sophisticated rotary laser levels will give both horizontal and vertical beam projection, but it must be remembered that these are accurate only if they are mounted on a stable structure.